WINGED ALIENS

WINGED ALIENS

A theory that that winged cryptid creatures are aliens and/or interdimensional beings

By Margie Kay

ISBN# 9780998855899

©Copyright 2021 by Margie Kay

Photos: Phillip Thomas, Margie Kay, Wikimedia Commons, and other named sources

Illustrations: Kristina McPheeters

Published by Un-X Media
PO Box 1166
Independence, MO 64051
816-833-1602
www.unxmedia.com

UNXMEDIA

PUBLISHING

DEDICATION

This book is dedicated to all the hard-working investigators of the world who work tirelessly to uncover the truth about unexplained phenomena.

ACKNOWLEDGMENTS

Thank you to Emmett Reary, Lon Strickler, Jeannette La Tulippe, Ken Gerhard, Sam Maranto, Bill Kousolous, Mindy Tautfest and others for sharing some of their reports.

Thank you to Joshua P. Warren for covering some of the Kansas City area sightings on his radio show on February 18, 2018.

Special thanks to all the witnesses who filed reports at Quest Paranormal Investigation Group and with me at margiekay.com and on my Facebook group Worldwide Mothman/Winged Creature Sightings

TABLE OF CONTENTS

INTRODUCTION

In the past 40 years doing paranormal investigations nothing has perplexed me as much as winged creatures. As a field investigator and Assistant State Director for MUFON, I have handled over 500 UFO reports for MUFON and have completed over 1,200 paranormal, and UFO reports on my own. I had taken several reports of giant birds prior to 2016, but that year things really stepped up and I received many calls and emails about sightings. I was not only receiving reports from complete strangers who found my website online, but from many of my friends and even family members. This piqued my interest, so I decided to start cataloging these sightings on my blog at www.un-xnews.blogspot.com in chronological order. As soon as I posted the article, more reports started to pour in.

Combined with reports of these creatures I found on other websites and received from other investigators - the numbers of sightings are compelling.

Some reports were few and far between, but in 2019 and 2020 I took at least 30 reports from witnesses who saw bird-like creatures with wingspans measuring 10' to 36' in width. Then in February of 2021 reports of tiny humanoids with wings began to surface.

Terms the witnesses used to describe what they saw include:

Mothman
Winged Humanoid
Pterosaur/Pterodactyl
Bat-Man
Thunderbird
Winged-Creature
Butterfly People
Batsquatch
Alien
UFO
Mimic
Fairy

It is possible that people are seeing the same type of creature but call it by different names. It is also possible that the creatures seen were in different stages of life such as youth or adult and could appear different yet be the same species. And we cannot ignore the possibility that there are multiple species involved, or that these creatures may be inter-dimensional beings.

Some people have asked if the large humanoid creature with wings is the Mothman and if it is a portent of a coming disaster. Of course, I do not know the answer to that question, but people who know about the Point Pleasant bridge disaster in West Virginia in 1967 and the Mothman sightings for several months prior may associate seeing the Mothman with a disaster due to that event and others. This case was covered extensively by John Keel in his book "The Mothman Prophecies."

There were reported sightings of a Mothman before other disasters such as the 1985 earthquake in Mexico City; the Chernobyl Nuclear disaster in 1986; the September 11, 2001 Twin Towers attack in New York City; and the Minneapolis Saint Anthony Falls Bridge collapse in 2007. There was even a reported sighting at the Fukushima power plant in Japan before the earthquake, tsunami, and meltdown of the nuclear reactor in March of 2011. Marcus Pules and his friend witnessed a Mothman circling the plant and sitting on top of a nearby building a few days before the earthquake.

There have been many sightings in the Midwest and other regions in the last few years. The one disaster we are all experiencing right now is the COVID Pandemic, so we must wonder if there is not some type of connection. However, these creatures have been seen at least since the early 1900's in many locations, so there may be no connection to disasters at all.

In speaking to numerous witnesses, I found that they were all hesitant to speak about what they saw, but upon finding out that I am investigating these strange creatures, contacted me anyway. Most just wanted someone to hear their story, and all seem very credible and authentic.

Sizes reported vary from wingspans of six feet up to 36 feet, prompting one to speculate about possible different types and ages of the creatures.

There are many questions that to date remain unanswered. What are the very odd, winged creatures? Are they some type of cryptid from Earth, or alien creatures from elsewhere? And is there more than one type? Where does it live and get food and water? And finally, what is it doing here and how does it remain so elusive?

Following is a compilation of reports taken by myself and short accounts from other investigator's websites and books in chronological order. Perhaps the reader will come to their own conclusion after reading these witness accounts.

HISTORIC CASES

1673 Alton, Illinois – Monster Bird Spotted by Explorers Jacques Marquette and Lois Joliet

Marquette wrote in his diary that their party saw a gigantic bird during their exploration and found multiple pictographs carved or painted on rocks and in caves along the Mississippi River between Illinois and Missouri. The Illini people called the creature the Piasa (pronounced PIE-a-saw), which means "bird that devours men." The Illini told Marquette that the bird had eaten dozens of tribal members, before the tribal warriors eventually killed the monster bird. According to Marquette the bird was "As large as a calf, with horns like a roebuck, red eyes, a beard like a tiger and a frightful countenance.

1869 OIL PAINTING OF JACQUES MARQUETTE [IMAGE: WIKI COMMONS/PUBLIC DOMAIN]

The face was something like that of a man, the body covered in scales, and the tail so long that it passed entirely around the body, over the head and between the legs, ending like a fish."

11

The image was visible until 1847 when the bluff was quarried away, but the legend endures. The image was recreated in recent years and is again on a bluff.

According to the Illini, "Having obtained a taste for human flesh, from that time he would prey on nothing else. He was artful as he was powerful and he would dart suddenly and unexpectedly upon an Indian, bear him off into one of the caves of the bluff and devour him. Hundreds of warriors attempted for years to destroy him, but without success. Whole villages were nearly depopulated, and consternation spread through all the tribes of the Illini. Such was the state of affairs when Ouatogo, the great chief of the Illini fasted in solitude for the space of a whole moon, and prayed to the Great Spirit that he would protect his children from the Piasa.

On the last night of the fast the Great Spirit appeared to Ouatogo in a dream and directed him to select 20 of his bravest warriors, each armed with a bow and poisoned arrows, and conceal them in a designated spot. Near the place of concealment another warrior was to stand in open view, as a victim for the Piasa, which they must shoot the instant he pounced on his prey. The next morning the warriors were quickly selected and placed in ambush as directed. Ouatogo offered himself as the victim. Placing himself in open view on the bluffs, he soon saw the Piasa perched on the cliff eying his prey. The Piasa rose into the air and darted down on his victim. Scarcely had the horrid creature reached his prey before every bow was sprung and every arrow sent into his body. The Piasa uttered a fearful scream, that sounded far over the opposite side of the river, and expired. Ouatogo was unharmed. There was wild rejoicing among the Illini and the brave chief was carried in triumph to the council house, where it was solemnly agreed that, in memory of the great event in their nation's history, the image of the Piasa should be engraved on the bluff."

Jon Russel continues in his report, "My curiosity was principally directed to the examination of a cave, connected with the tradition as one of those to which that bird had carried its human victims. Preceded by an intelligent guide, I set out on my excursion. The cave was extremely difficult of access and at one point I stood at an elevation of 150 feet on the perpendicular face of the bluff, with barely room to sustain one foot. After a long, perilous climb we reached the cave. The floor of the cavern throughout its extent was one mass of human bones. To what depth they extended I was unable to decide, but we dug to the depth of 3-4 feet in every part of the cavern, and still we found only bones. The remains of thousands must have been deposited here."

Sources: St. Louis Post Dispatch, October 29, 2017 and https://www.ancient-origins.net/myths-legends/tradition-piasa-and-mysterious-rock-art-mississippi-001452

The reader will find other accounts following which involve strange, winged creatures with human-like faces.

Could the legend of the Piasa have some truth to it? Was there a real creature that preyed on humans? Russel's discovery of the thousands of human bones in the cave is compelling evidence. Many reports of gigantic birds are from locations close to the Mississippi River, so perhaps there are some remaining Piasa in the region. (See the 1940's sightings later in this book)

Ancient Thunderbird petroglyphs in a Park in Missouri

Petroglyphs attributed to the ancient Cahokia people are carved into the rocks at Washington State Park in Missouri about 60 miles southwest of the Piasa image. This is located near DeSoto, Missouri and is open to the public. Visit mostateparks.com for more information.

According to Algonquian mythology, the thunderbird controls the upper world. The thunderbird creates not only thunder by flapping its wings, but lightning bolts, which it casts at the underworld creatures. (Note the similarity between this account and the Van Meter Visitor shooting lightning bolts mentioned later in this book).

Thunderbird drawings date to at least 1250 AD and have been found in many regions in the United States, Canada, and Central America. Perhaps the Thunderbird really exists and is not a legend after all. How else can you explain similar petroglyphs from many different regions?

Following are photos of petroglyphs from Missouri, The Great Lakes, and Wisconsin.

Winged Aliens

THUNDERBIRD PETROGLYPH PHOTO WASHINGTON SP PETROGLYPH DEC2011
N28.JPG WIKIMEDIA COMMONS

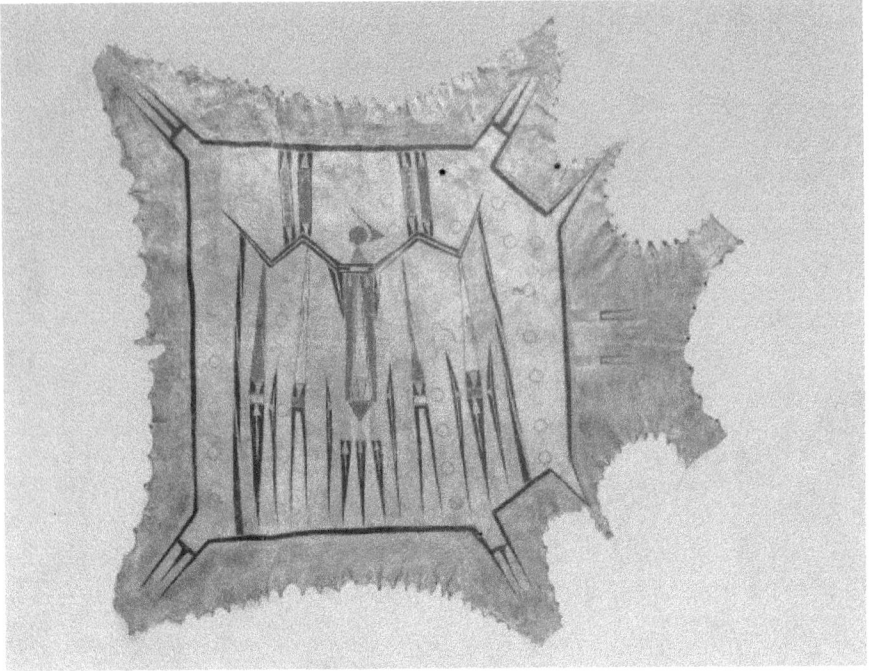

FIGURE 1 PAINTING OF A THUNDERBIRD FROM THE GREAT LAKES REGION

By musée du quai Branly, photo Patrick Gries - http://www.quaibranly.fr/fr/collections/collections-thematiques/collections-amerindiennes-de-nouvelle-france.html, Public Domain, https://commons.wikimedia.org/w/index.php?curid=34172342

Winged Aliens

THUNDERBIRDS CARVED IN SANDSTONE AT TIN BLUFF, JUNEAU COUNTY, WISCONSIN
BY PREHISTORIC PEOPLE
BY SIXA369 - OWN WORK, CC BY-SA 3.0,
HTTPS://COMMONS.WIKIMEDIA.ORG/W/INDEX.PHP?CURID=20076003

Unknown Location During the Civil War:

An authentic photograph surfaced a few years ago which shows a group of Civil War soldiers standing next to the carcass of a Pteranaodon. The photograph is real, but the creature remains unidentified. However, it looks like a pterosaur. It is unclear whether the soldiers shot the creature down or if it was found dead. A second, fake photo has also been circulating the internet. In the second photo modern men recreated the original scene for a documentary film.

AUTHENTIC CIVIL WAR PHOTOGRAPH OF A PTEROSAUR
PHOTOGRAPHER UNKNOWN

It is a fact that Pteranodons existed at one time. According to David Hatcher Childress (http://s8int.com/dino21.html) The largest fossil ever discovered was at Big Bend national Park in Texas in 1975. It had a wingspan of 51 feet. Other pterodactyls found were much smaller, with wingspans measuring from 8 to 20 feet. One possible explanation for modern sightings of creatures that fit this description is that somehow pterosaurs survive to this day. Childress believes that MothMan and giant eagle sightings can be attributed to these living flying Pteranodons.

There are a number of reports following which substantiate the theory that some of these species are alive and well yet today.

PHOTO OF A GIGANTIC BIRD, ORIGIN UNKNOWN, APPEARS TO BE CIVIL WAR PERIOD.

PHOTO OF UNKNOWN ORIGIN SOURCE: HTTPS://THEMOTHMAN.FANDOM.COM

1868 Tippah County, Missouri – Eight-Year-Old Boy Snatched up by Giant Bird

According to author Troy Taylor, Jemmie Kenney was picked up by a gigantic "eagle" as witnessed by his teacher. The bird dropped the boy, but he later died from injuries sustained in the fall and the talon injuries on his shoulders. According to zoologist Dr. Bernard Heuvelmans, even the largest eagle could not possibly carry off a child, therefore, it must have been some other type of bird.

April 26, 1890 – Giant Bird in Tombstone, Arizona

The Tombstone Epitaph printed an article titled "Found in the Desert. A Strange Winged Monster Discovered and Killed on the Huachuca Desert." In this article the writer describes "A winged monster, resembling a huge alligator with an extremely elongated tail and an immense pair of wings" was found by two men on horseback, who subsequently shot and killed it after chasing it for some time.

The men reported that the head was about eight feet long, the wings were joined with the body, the jaws were thickly set with strong sharp teeth. The eyes were as large as dinner plate and protruded about halfway from the head. The wings measured 160 feet from tip to tip. The wings were composed of a thick and nearly transparent membrane and there were no feathers on the wings or body. The finder, who is not named in the article, stated that he planned to skin the creature and send the hide east for examination by the eminent scientists of the day (the Smithsonian, perhaps?).

TOMBSTONE EPITAPH APRIL 26, 1890

The men reported that the head was about eight feet long, the wings were joined with the body, the jaws were thickly set with strong sharp teeth. The eyes were as large as dinner plate and protruded about halfway from the head. The wings measured 160 feet from tip to tip. The wings were composed of a thick and nearly transparent membrane and there were no feathers on the wings or body. The finder, who is not named in the article, stated that he planned to skin the creature and send the hide east for examination by the eminent scientists of the day (the Smithsonian, perhaps?).

This story gets even stranger due to the fact that many people swear they saw a photograph of this creature hanging up with several cowboys standing next to it with their arms outstretched. It is said that the photo was published in National Geographic, FATE Magazine, Saga, true, and even on a Canadian TV show called The Pierre Benton Show. John Keel and Ivan T. Sanderson are among those that claim to have seen that photo, as I did in the 1960's in one of my grandparent's magazines. Yet, to date, no photo of this creature can be found. How is it possible that so many of us saw that photo, yet it no longer exists anywhere? This is bizarre, to say the least.

October 1903 Van Meter, Iowa – The Van Meter Visitor
Many reports were made of sightings of a creature that was described as half-man and half-animal with large bat-like wings, brilliant beams of light that shot from its forehead, and a horrid foul aroma that it emitted during the week-long sightings.

U.G. Griffith had the first encounter with the creatures at 1:00 am. He thought a burglar was on one of the rooftops of a downtown building, but as he approached the burglar jumped to another rooftop across the street and disappeared. The distance would have been impossible for a man to reach.

The Second encounter occurred at 1:00 am the next day when Dr. Alcott woke up with a bright light shining into his window. The doctor grabbed his gun and ran outside to confront what he suspected was a burglar, only to find a tall bipedal humanoid creature with bat-like wings. The source of the light was coming from a blunt horn in the creature's forehead. He opened fire with his gun, but the five shots had no effect on the creature, so he retreated inside his building for safety.

A third encounter occurred on the third night when at 1:00 am Clarence Dunn, the local banker had a run-in with the creature at the bank where he was spending the night to protect it from the "burglar." A bright beam of light shone inside the front window and moved from side to side until it focused on Mr. Dunn. The bank then fired his shotgun through the window, but the light simply disappeared. The next morning Mr. Dun found three-toed footprints in the ground and made plaster casts of them, which later disappeared.

That same evening yet another encounter occurred at the Van Meter hardware store. Owner O.V White awoke to a metallic rasping sound outside his room on the second floor. He went to the window and saw a creature sitting on a telephone pole in the rain. Mr. White fired and shot the creature, which remained unfazed, but released a horrible smell that knocked the man unconscious. White's neighbor woke up to the noise and saw the creature coming down the telephone pole. His descent resembled that of a parrot using its beak. After getting to the ground the creature stood up and Mr. Gregg could see that it appeared to be at least eight feet tall. The creature then again shone the bright light from its forehead to look around, then headed to the old coal mine.

The next day the townspeople were talking about the creature sightings. That night they heard eerie sounds coming from an abandoned coal mine. The sound was "as though Satan and a regiment of imps were coming froth for a battle" as described in the Des Moines Daily news October 3, 1903.

Townspeople approached the mine and saw not one, but two such creatures emerge from the mind and fly off. A mob collected at the mine and awaited the creatures return, and when they did so before dawn the men fired at the creatures with their guns, again to no avail as the bullets seemed to have no ill effects. The creatures again released a horrid smell and went into the mine, which was 257 feet deep. Finally, on October 3, 1903 men barricaded the mine entrance, sealing the creatures inside. The Van Meter Visitors were never seen again – until the 1980's and later (see following reports).

What could these creatures have been? Certainly, it was not a hoax as many people saw them and had nothing to benefit from reporting it, and the town never received any profit from the reports. The story only came to light in 2013 when "The Van Meter Visitor (A True & Mysterious Encounter with the Unknown)" was released by Chad Lewis, Noah Voss, and Kevin Leen Nelson. In the book, the authors state that the Ioway tribe that lived in the region had an oral history of humanoid Thunderbirds that shot "lighting" from their eyes.

It sounds as though this was a real event and the creatures did (or do) exist.

Van Meter hosts the Van Meter Visitor Festival every year.

According to Chad Lewis a huge bat-like creature has been seen in the area in recent years.

Sources: The Van Meter Visitor: A true and Mysterious Encounter with the Unknown by Chad Lewis, Kevin Lee Nelson, and Noah Voss; the bigfootdiaries.blogspot.com; cryptids.wikia.com

VAN METER HOT UNDER THE COLLAR

TOWN HAS BEEN MALIGNED BY GHOST STORIES.

Citizens of the Place Feel Indignant Over the Matter, as It Gives the Place an Unenvi-able Reputation.

The town of Van Meter is justly indignant over a series of articles that have appeared in the Daily News, and the Capital is in receipt of a number of letters from citizens of that place who feel highly indignant over the matter. The articles alleged that the town was highly wrought up over the alleged affair. The principal article started out with the following:

"Quite frequently one hears of a haunted house, but for a whole town to have 'em, is a different proposition. Van Meter, a town of about 500 souls, lying 30 miles west of Des Moines, alone enjoys the distinction of being haunted. Queer noises are heard, hideous apparitions have been seen and weirdly lights move around in a mysterious manner."

VAN METER NEWSPAPER ARTICLE

26

1903 - 1999 SIGHTINGS

The following cases were reported to me directly via Facebook, email, in person or by phone, except for those noted from other sources. I did not include the more than 125 sightings of giant flying humanoids in and near the Chicago area since Lon Strickler and others have covered those sightings extensively, and I did not include most of the Point Pleasant, West Virginia sightings. Most of the sighting reports covered in this book are from the State of Missouri, with a few from surrounding states. In doing extensive research online, it appears that as far as the Midwest region goes, Missouri is a hot spot for such creatures.

One thing is certain – a lot of people have seen strange large creatures with wings. As far as credibility goes many of these reports come from persons known to me and I can vouch that they would not make up such a wild tale, and as for others I can say that they sounded credible and sincere and did not seek publicity.

1919/1920 Camden, Missouri- Large Bird with a Human Face

A friend and local retired radio producer Vern Windsor reported that his 8th-grade teacher told him that when she was a child in 1919, she and her family saw a gigantic bird with a human face circle them very close, scream at them, then fly off towards the hills. The following is a statement from Vern:

"Margie, I remember a story my 8th grade teacher told the class about a sighting she had. This would have been around 1919 or 1920. She lived in Camden Missouri at this time. When she was a preteen, her family was walking to church when a large, winged thing with a human-like head swooped in front of her family, screamed at them, flew around them for a minute and then flew off to the north. She said they walked a little down the road and her mom asked her dad if he saw that thing. Her dad answered her with (paraphrasing here) "don't talk about that! We didn't see anything!" She told us the family tried once a few years later to talk about the flying thing and her dad got mad at them, yelling the same things."

Camden is a small town in **Ray County, Missouri**. The population was 191 at the 2010 census. Camden was a Missouri River boat stop until the early 1900s and the Missouri River's Camden bend was cutoff after major river flooding, moving the new channel south.

Many sightings have occurred near the Missouri River. If the creatures are using the river as a food and water source it would make sense that it would occasionally be seen by people.

April 4, 1948 Alton, Illinois- Thunderbird Sightings by Many Witnesses
There was a rash of sightings of huge birds in and around Alton, Illinois not too far from the ancient Piasa Thunderbird

drawing (see the 1977 report following). On April 4th, 1948, an Army Colonel named Walter F. Siegmund said that he saw one of these creatures as he was speaking with a local farmer,

"I thought there was something wrong with my eyesight. But it was most definitely a bird, and not a glider or a jet plane. It appeared to be flying Northeast… and from the movements of the object, and its size, I figured it had to be a bird of tremendous size."

April 24, 1948 Alton, Illinois- Thunderbird

A few days after the April 4 sighting there was another sighting in the skies above Alton. This time, many people saw them.

Flight instructors at the local airport said the largest of the birds cast a shadow, "as big as a Piper Cub from 500 feet up." Other eyewitnesses maintained that the birds were as large as small airplanes, and far bigger than any eagle they had ever seen.

Alton is a town on the Mississippi River located 18 miles north of St. Louis, Missouri.

April 1948 St. Louis, Missouri – Huge Bird Spotted by Many Witnesses

At least 13 witnesses reported seeing a gigantic, grayish-black bird with a torpedo-like body soaring in the skies above South St. Louis, Alton, Glendale, Overland, and Richmond Heights. People described the creature as "immense," "incredible," and "enormous." A police patrolman, and a corporal saw it as well. Officer Francis Hennelly said that "Its wings were flapping, and it was headed southwest, flying at an altitude of several hundred feet. I thought it was a large eagle, but I've never seen one that big before." One witness said that unlike normal birds, this creature flapped one wing, then the other. Another witness watched as the giant bird flew towards the Mississippi bluffs, where he surmised would make a good habitat for such a creature.

The original story posted in the St. Louis Post-Dispatch was titled "The Gigantic Bird Mystery: That Thing in the Sky- is it Really a Bird or is it a Plane or a Witch?

Read the original 1948 article here: https://www.stltoday.com/news/multimedia/graphics/the-giant-bird-mystery/pdf_a594d89f-70c1-5a4e-be0d-5d89366fec44.html#tracking-source=article-related-bottom
Source: St. Louis Post Dispatch, October 29, 2017

1954 Kansas City, Missouri – Fairy Ring and Fort
When James Bair was a child he was informed by friends and neighbors that there was a fairy ring and fort on the UMKC Campus. These are located on the west hillside of the Miniature and Toy Museum approximately 35' from the parking lot. James and his wife, Margaret, heard tales about people who have seen fairies at this location and have visited the site on several occasions, although neither has seen the fairies themselves. Mushrooms often grow in a perfect circle in this ring and people are warned not to walk on the fairy fort.

1955-56 Kansas City, Missouri- Giant Birds Sighted by Multiple Witnesses

In locations near 47th & Denver close to the limestone quarry and KC Zoo/Swope Park area multiple sightings of large, winged creatures were seen by a number of persons over a two-year period. James Bair is a retired schoolteacher and Field Investigator and section director for Missouri MUFON. James and his family and neighbors saw these winged creatures over a two-year span. James was eight and nine years old at the time. In one instance his father called him outside and said excitedly, "Take a look at that great big eagle on top of the garage!" Then, he thought better of it and said it was a huge owl. But neither of them knew what it actually was, and it scared them because it was just too big to be a known bird. The creatures had feathers, a white head with oval egg - shaped black bodies.

James and his friends used to play along Brush Creek and would sometimes see these big birds. He believes that some of the creatures were short and stocky, while others were taller and thinner. None had a long tail. They also struggled to get off the ground when starting to fly. In the 50's eagles were in decline, so it was not common to see eagles at that time.

James said that he saw several of these creatures standing next to the banks of Brush Creek. "They stood about 5' 8" tall, about the same height as my dad, and they never made a sound," James said. The neighbors and other family members saw these unknown creatures on multiple occasions, sometimes flying in the air, sitting in trees, or standing on the ground on two legs. James would be super - vigilant when he and his friends went to the creek, keeping watch for these large creatures, which scared him.

The birds seemed to make appearances after dusk but not during the day. James surmises that they may have been hiding out in the caves at the quarry during the day, and came out at night to eat, presumably fish from the creek or chickens and other animals that many people had in those days. James and his family lived between Brush Creek and the Blue River. The Blue river runs past mines and a police pistol range. It was undeveloped in that area and was mostly forest, where any type of creature could live. Today James says that he could not rule out that the creatures were pterodactyls.

BRUSH CREEK FLOWING THROUGH KANSAS CITY, MO. PHOTO RUTH MARSHALL, U.S. ARMY CORP OF ENGINEERS. PUBLIC DOMAIN

The sightings mysteriously stopped, and the birds were not seen by the group of friends again. James Bair has lived in the same area his entire life and has not seen these creatures since 1956.

I have taken a number of reports of sightings of giant winged humanoids in and around the Swope Park area, which is close to this site. Swope Park is an area where any type of animal could easily hide (see further information about the area in my conclusion).

Source: James Bair, Field Investigator and SW Section Director for Missouri MUFON.

1957 Braidwood, Illinois – Winged Humanoid

From the Singular Fortean Society: 79-year-old Gerald Turrise reported that in 1957 when he was hunting in the woods for pheasants and rabbits with his brother-in-law and brother, they came across a very strange creature. As they stood next to a large tree in the center of a field in mid-morning a huge man-sized creature sailed over our heads into the woods across the road. "We were stunned, we just looked at each other too dumbfounded to speak." The witness described the creature as having the body of a large man with legs but covered all over with dark tan colored feathers and with large wings. This incident is the oldest known sighting of a Mothman like creature in the Lake Michigan area. Note: The witness observed a UFO in 1963 with stationed at a Nike Hercules missile site in Northfield, Illinois. He was contacted by Dr. J. Allen Hynek about the incident.

1966 - 2021 Point Pleasant, West Virginia – Mothman Sightings

This is probably the most famous Mothman sighting case ever recorded. John Keel wrote "The Mothman Prophecies" in 1975, which covers the strange events which occurred for months prior to the Silver Bridge Collapse, and still continue today. Witnesses in 1966 saw UFOs, strange lights, had visits from unfriendly Men in Black, became psychic, and saw a menacing human-like creature with wings. The first reported Mothman sighting was on November 12, 1966 when two gravediggers reported seeing a winged humanoid glide over their heads. Three days later, two couples saw a massive humanoid with a 10-foot wingspan and glowing eyes. The creature chased their vehicle as they drove down the road. More people came forward with reported sightings after that and there were over one hundred sightings reported between 1966 and 1968.

Many thought that the Mothman was related to the Silver Bridge disaster in 1967. The bridge connected Point Pleasant, West Virginia to Galliopolis, Ohio and resulted in the tragic deaths of 46 people.

A movie called the "Mothman Prophesies" was made in 2002 starring Richard Gear which covered the strange events. Keel's conclusion is that there was something supernatural going on in the area. The sightings are numerous and have continued throughout the years. In John Keel and Andrew Colvin's 2015 book "The Book of Mothman," the authors relate a case that is of particular interest when trying to establish a link between UFOs and winged humanoids: At 10:30 p.m. on May 19, 1967 two women driving on Route 62 heading north from Pont Pleasant said they saw a dark form with two bright red lights on it circling a tree. The winged creature was bigger than a man. Suddenly a larger red light appeared and moved towards the dark circling

MOTHMAN STATUE IN POINT PLEASANT, WEST VIRGINIA IMAGE: GOOGLE EARTH

figure. The two merged then flew off. The two women believed that they had just seen Mothman rendezvous with a UFO.

There is another strange fact about the Braxton County area, home of the so-called "Flatwoods Monster," according to the Keel and Colvin in the past three years prior to 2015 no less than 20 teenaged boys have vanished suddenly without a trace. Could this be related to the winged creatures?

Be sure to visit the Mothman Museum (www.mothman museum.com) or attend the annual Mothman festival in Point Pleasant.

1966 -1968 Galliopolis, Ohio – Mothman

Residents reported hundreds of sightings of the Mothman at the same time people in Point Pleasant were seeing the creature. Gallipolis is located across the river from Point Pleasant.

Spring 1966 Kansas City, Missouri – Fairies inside a House

This is my own sighting. My younger sister and I had just gone to bed one evening at our grandparent's house. We were aged 10 and 8. It was dusk and there was a small amount of light outside. The two windows were open, and the screens were closed. I was shocked to see two 6" tall fairies fly into the window at the front of the house. The dark-haired fairy flew over to me, and the blond-haired fairy flew over to my sister. The tiny creatures hovered approximately 12" or less from our faces while we stared, transfixed at them. They both looked like humanoid females with wings. Then, as quickly as they came into the house, they exited through the window at the side of the house. My sister and I jumped up and both said at the same time "Did you see THAT?" We were perplexed as to how these creatures could possibly exist, not to mention the fact that they flew through screens!

Now that we had seen fairies and knew for a fact that they existed we began to look for them in my grandmother's garden outside. One day, I saw several fairies flying around the Lilly's of the Valley. I ran inside to tell my grandmother and mother about this and they both said they knew about the fairies and that they liked to hang out in that particular spot, which was located under the big kitchen window and near the door to the screened - in porch. In fact, my grandmother's favorite flower was Lilly of the Valley, and she grew them purposely to attract fairies. I remember seeing the fairies in this spot only a few times as a child. Later in life, there would be more to the story (see my 1984 report). It should be noted that my family has experienced many strange things including UFOs, spirits, time anomalies, and more.

Summer 1968 Keeneyville, Illinois – Thunderbird Frightens Boy

David T. St. Albans wrote about his own sighting in his 2012 Blog. David was staying with his father in a rural area outside of Chicago among a quarry and corn fields. The two-acre lawn had tall grass that needed cutting and David was in the process of doing that when he spotted a huge bird that oddly seemed to be going in and out of time or dimensionally shifting. David noted that the wingspread was at least eight to ten feet. The bird appeared to be a vulture-like and was brownish black with a bald head and long, extended beak which was more like an Albatross beak. It had sharp white teeth, which made it look even more frightening.

Source: https://whisperindave.wordpress.com

1974 Kansas City, Kansas – Pterosaur Sighting

At 8:00 am one morning one day in 1974 witnesses observed a flying bird that could only be described as a pterodactyl. The witness said that the winged creature had a long head and huge leathery wings with a 15-foot wingspan. As it flew overhead, several people watched in awe without saying a word until it was out of sight.

1976 Rio Grande Valley, Texas – Big Bird Startles Texans

People started seeing gigantic birds in the Rio Grande Valley area beginning in January 1976 when Tracey Lawson and her cousin Jackie Davies saw a creature they could not imagine existed. They said that it was over five feet in height, had dark ready eyes, a bald head, gorilla-like face and a long beak. Their father found odd three-toed tracks pressed about one inch into the ground, indicating that something heavy made the tracks.

On January 14 Armando Grimaldo heard a sound like a bat's wings flapping and turned when a gigantic bird grabbed him with its talons and tore his shirt. Armando described the creature as having a monkey-like face and red eyes, but no beak.

Several schoolteachers saw the creature and estimated it had a wingspan of at least 15 to 25 feet. After doing some research, they decided the bird must have been a pterosaur. The incidents were covered in several newspapers.

'Big Bird' Tale Winging Along

SAN ANTONIO, Tex. (UPI) — South Texas has its own private unidentified flying object — a huge winged creature that glides silently through the streets at night terrifying citizens and police.

For want of a better name, residents are calling it "Big Bird."

Rumors of the bird — described as having a wing span the length of an automobile — began on a school playground in Robstown, Tex., two months ago.

Reports from persons who claimed they saw the huge bird led to tongue-in-cheek spoofs by a Corpus Christi television station and newspaper which were thought to have put the legend to rest.

They didn't.

Numerous sightings of big bird continue to come in from miles apart, especially in the Lower Rio Grande Valley along the Mexican border.

Last week a Harlingen television station showed giant bird prints it photographed in a freshly plowed field in the area of big bird sightings. The prints showed a three-toed impression nine inches wide and 12 inches long.

The same day San Benito policemen Arturo Padilla and Homero Galvan, traveling in separate squad cars through dark streets, reported seeing a huge bird with a 15-foot wing span gliding through the Valley city, without a single flap of its wings.

"It more or less looked like a stork or pelican type of bird," Padilla said. "The wing span I guess was about like a pretty good sized car, about 15 feet or so. The color was white.

s Block

SPRINGFIELD MO DAILY NEWS JAN 13, 1976

Summer, 1976 Idaho – Pterodactyl Sighting

A woman and her father saw a huge Pterodactyl on the side of the road near the Deer Flat National Wildlife Refuge not far from Boise, Idaho. It was late afternoon, and they got a good look at the creature, who was standing still then took flight.

Melinda Buress Fauth said "My dad and I were driving out in the desert part in the countryside on our way home from Boise area. My dad was the one driving. We noticed in the distance ahead to the left of us what appeared to be a man standing out in an already harvested field. We talked wondering what he was doing out there. As we kept driving closer along the crop

40

DRAWING OF CREATURE BY MELINDA FAUTH

fence line whatever it was spread wings totally catching us off guard. It proceeded to flap its wings and started low across the field to build up speed. We got closer and it got closer. We did not know if it was going to get up enough height and thought we might actually hit with our vehicle. We were headed west, and it was headed more north. We were pretty close to it when it lifted up and proceeded to miss us and flew right over and above us, I would say less than 20 feet above us. We got a good close-up view of it. The body was about the height of a man or a door. It had a short curved downwards beak with a short crest on the back of its head and a short neck and its head looked like a bird and not a mammal. The wings resembled bat wings with a short tail. If it had feathers it was so dark, you could not make any out. It was solid black body. One wing was like the height of a ceiling or more. Total wingspan I would roughly estimate at least 16 feet to 20 feet. I am using measurements inside of a regular sized home for size comparison. I always called it a pterodactyl because it looked like something out of a dinosaur book. It was headed towards the Cascade Mountains on the other side of Boise, Idaho. It kept flying higher and higher until it was out of sight."

Melinda said that at the time the area was not built up much and there was a lot of open desert with no houses around where they were driving.

July 1977 Lawndale, Illinois- Giant Bird Tries to Carry off a Child

This is probably the most well-known case of a giant bird sighting that has ever been published. Multiple newspapers carried the story which was reported by several credible witnesses, and it has been presented in television documentaries.

From the Boston Evening Globe July 28,1977 edition:

CARRIED OFF

"10-year-old Marlan Lowe and his mother Mrs. Ruth Lowe claim that one of two large black birds with eight-foot wingspans tried to carry Marlan off in its claws Monday evening in Lawndale, Illinois. Although several bird experts say that no bird native to Illinois could lift 56-pound Marlan, Mrs. Lowe says that Marlan was carried over 35 feet before the bird dropped him when she struck the bird with her hand." (UPI)

In an article by Jerry D. Coleman of Crytozoology.com Mrs. Lowe made a statement to the police, Ruth described the bird:

"It had a white ring around its half foot long neck. The rest of the body was very black. The bird's bill was six inches in length and hooked at the end. The claws on the feet were arranged with three in front, one in the back. Each wing, less the body, was four feet at the very least. The entire length of the bird's body, from beak to tail feather was approximately four- and one-half feet."

According to the Mt. Vernon Register-News August 4, 1977 sightings of giant child-snatching birds were reported by several persons. According to the article "Texas John Huffer, and employee of a Champaign television station, reported seeing two giant birds on Saturday and said they had droppings the size of baseballs." The article discredited the sighting reports, stating that they were no more fearsome that the lowly turkey vulture according to bird experts.

Marlon Lowe had difficulty coping with the ordeal and the media attention. His hair turned gray for a time until it grew out. Neighbors harassed the family, even leaving dead birds on their doorstep.

The giant winged creature sightings were discredited by Jack Ellis, supervisor of wildlife resource for the state conservation department He said the creatures were likely Turkey vultures, which have a wingspan of up to six feet.

It is interesting to note that the Cahokia tribe lore includes giant Thunderbirds. One of the famous cliff drawings is known as the Piasa Bird located outside of Alton, Illinois. It was first discovered in 1673 by Jacque Marquette. The Cahokia tribe calls these creatures Thunderbirds after the sound they make when they flap their gigantic wings.

In the 1940's there was a large number of sightings of huge birds in and around Alton (see the previous 1940 report).

Perhaps these are not mythical creatures after all.

The Piasa Thunderbird photo:Wikimedia Commons

July 28, 1977 McLean County, Illinois – Multiple Gigantic Bird Sightings

According to vistcryptoville.com farmer Stanley Thompson spotted a large bird flying over his farm. The bird had been seen by the family just three days prior while they watched radio-controlled airplanes. The creature flew close the model planes, and it dwarfed the small planes. Mr. Thompson said he thought that the bird's body was about six feet long and had a wingspan of nine to ten feet. County Sheriff Sergeant Robert Boyd received several reports of giant bird sightings but only investigated this one because he knew that Mr. Thompson was credible.

On the same day mail truck driver James Majors saw two giant birds near Bloomington, Illinois as he drove from Armington to Delevan. Both birds had 10-foot wingspans. Each animal appeared to fly down into fields and pick up small animals.

Also, on the dame day, Lisa Montgomery of Tremont saw a huge bird with an estimated wingspan of seven feet fly overhead as she washed her car. The creature flew towards Pekin.

July 30, 1977 More Huge Birds Reported throughout the day from north central to south central Illinois

A huge bird was sighed in Downs, Waynesville, Decatur, Macon, and Sullivan. One witness obtained video of the bird that he later sold to a Champaign television station. He estimated the bird's wingspan to be over 12 feet.

July 31, 1977 Giant Bird in Bloomington, Illinois

Mrs. Albert Dunham saw a large bird with a white ring around its neck as she looked out her second story window. Her son followed the bird to a nearby landfill, where he lost sight of it.

July 1977 Illinois -Thunderbird Captured on Film!

A man by the name of Chief A-J posted a video on YouTube that was shot by him in 1977. Many reports of giant living thunderbirds were pouring into radio and TV stations across central Illinois. CBS hired Chief A-J of the Central Tribal Native American Council to capture these creatures on film. Chief A-J was trained by the U.S. Marine Corp as a combat photographer so was well - suited for the task. He decided to cruise the shoreline of Lake Shelbyville by canoe with his 16mm news camera in hand and was able to obtain footage of two of the gigantic birds. This video is posted at tinyurl.com/2kuynk4z

Credit: Chief A-J, www.chiefaj.com

August 11, 1977 Giant Prehistoric Bird Lands in Tree in Odin, Illinois

Jon and Wanda Chappell saw a grey, black bird with an estimated wingspan of 12 feet land in a tree near their home. They described it as looking "prehistoric," and capable of carrying of their small daughter if it had the chance. The couple hesitated to report their sighting for fear of repercussions.

1977 near Herrick, Illinois – Huge Birds Spotted by Multiple Witnesses

The witness estimated the wingspans to be at least 10 feet. People in Illinois stopped reporting seeing these birds, although they continued to have sightings for several years, at least until 2002.

Giant bird sightings in Illinois 1977-1978
Image: Google Earth/Margie Kay

1982 Alliance, Ohio – Mothman

Several friends were leaving a party late one evening just outside Alliance, Ohio. The witness noticed red lights in their rearview mirror and assumed it was a car, but it moved up to quickly. At that point, the friends realized that it was actually some type of creature with red glowing eyes. It stopped about 10 feet away from the vehicle and the group got a better look at it. It had a vague humanoid shape.

Source: exemplore.com

1980's Canton and Ravenna, Ohio – Mothman

A 12-year old boy was frightened by a winged transparent creature with red burning eyes which appeared in his bedroom. The creature appeared to be made up of tendrils of a black, smoke-like substance. It hovered over the boy, then laughed in a deep, echoing male voice. The boy remembers being paralyzed and unable to move.

The second encounter occurred when the young man was age 16 and living in Ravenna, Ohio. The second event played out nearly exactly as the first, but it lasted for about an hour. He called the creature the "boogieman" as a child, but later realized after reading "The Mothman Prophecies" that it was more likely a Mothman.

Source: exemplore.com

Some people believe that in cases where the witness is unable to move, they must have a condition called sleep paralysis. I am of the opinion that this is not always the case and that extraterrestrials and other interdimensional beings have the ability to make people paralyzed. With the description of the creature being transparent, this leads me to believe that it is an inter-dimensional creature.

1984 Batman/Mothman Creature Sighted at Monkey Mountain Nature Preserve in Missouri

Two people who were visiting Monkey Mountain Nature Preserve near Oak Grove, Missouri and saw an unusual creature while hiking there in the summer of 1984. Terre Tweedie and her husband were walking along a path in the afternoon when they both saw a seven-foot plus tall humanoid/bat-like creature with exceptionally large wings standing next to a tree. The creature fluttered its wings as the couple approached on a trail, stopping them dead in their tracks. It was near dusk, but the witnesses could see the creature clearly.

"It had a bat-like face and a man-like body," said Terre. "The wings were not spread out, but you could tell they were huge. The first thing I thought was that it could have been the Mothman. It had glowing red eyes, and that was the scariest part of the sighting."

Terrified by what they saw, Terre and her husband slowly backed up on the trail, then turned and walked back down the trail, not looking back. Terre says she will not go back to Monkey Mountain.

It should be noted that a lot of strange events have occurred at Monkey Mountain, including multiple sightings of Sasquatch and a Chupacabra like animal, and people have heard strange noises and animal calls that cannot be identified, along with a general eerie feeling of being watched. One woman told me "As I walked into an area on Monkey Mountain suddenly all sounds stopped. It was so eerie that I left immediately because I knew something was not right." Many people who have visited the area will not go back.

Monkey Mountain Nature Preserve Photo:Wikimedia Commons

In the 1970's a group of people recorded sounds at Monkey Mountain and later had them analyzed in the 2000's by language expert Scott Nelson, who determined that the calls were not from a human being or any known animal. Nelson played the tape at a 2014 conference in Kansas City, and I can verify that the sounds were very strange. Scott Nelson believes that the sounds came from a bigfoot.

A lot of people wonder how Monkey Mountain got its name. There is a rumor that a truck load of monkeys escaped from a vehicle carrying them from a zoo in the 1920's and they escaped into the woods. However, I can find no newspaper articles that would support this tale. Later, due to the Sasquatch sightings and strange sounds in the area, the name took on a different meaning altogether.

June 1984 Ireland – Fairies

Margaret Bair lived in Ireland in the 1980's and she frequently heard about fairies. Once she was walking down the road with a group of people on a tour and wanted to cut through a neighbor's land as a shortcut. He told the group to keep away from a certain area that was known to be a fairy fort. There were several small hills that were fairy land. It seems that the fairies do not like people to go into their territory, and they would cause problems for people who did so. Margaret veered away from it and always remembered to walk around the fort from then on. The landowner was very serious about this and it was not a joke.

Spring 1984 Kansas City, Missouri – Fairy Sighting and a Missing Child

On a Spring evening in 1984 when my daughters were age 8 and 6, they suddenly came running into the kitchen where I was sitting with my grandmother. They had just been sent to bed a few minutes prior. What they had to say was shocking to say the least – they both said they saw a fairy come into the front window and go out the side window – exactly as my sister and I experienced many years earlier. None of us had ever mentioned the prior event to my children, so this was shocking for all of us.

Not long after this, another event occurred that had us concerned. I was outside in the yard with my grandmother working in the garden while my two daughters ran circles around the house. This was a favorite game of theirs, with the youngest chasing the oldest. At one point, my youngest daughter came running around the corner and stopped dead in her tracks. She said, "Where is my sister?" Her sister should have been just a few feet in front of her. The spot where she stopped was right in front of the kitchen window. We were perplexed and thought that my daughter must have been hiding somewhere, but it would have been impossible. We looked outside and inside, and at the neighbor's house for 45 minutes. Just as I was getting ready to call the police, my oldest daughter came running around the corner and stopped short. She said, "Where is my sister?" She was surprised that her sister was no longer following her and that my grandmother and I were not in the same location. We surmised that my older daughter had been taken by the fairies for 45 minutes, then returned. After that, I would not allow anyone to go near the fairy spot, and we all gave it a wide berth when walking by it.

I should note that many strange events occurred at this house and over the years my family has decided that there is some type of portal in that location.

1985 St. Louis, Missouri -Pterosaur

As reported by Jonathan David Witcomb in his book and blog "Live Pterosaurs in America," a witness, who was age 10 at the time, and her father were at a family barbeque when they heard a horrific screech and looked up to see a real live Pterosaur that was the size of a small aircraft. The wings were like bat wings, and it had a long beak with razor-line white teeth. It had a long tail with a tuft like shape at the end, but no fur or feathers. They got a good look at the creature because it was only about 50 feet above them. The parents grabbed their children and ran inside the house. This story received a lot of media attention. See the full article here: https://www.livepterosaurs.com/inamerica/blog/?p=1543

Early 1990's Maryville, Missouri- Huge Bird Stuck in the Mud

A family was out hunting for quail and pheasant when ad midday their dog started barking wildly. The looked where the dog was pointing and saw a huge, monstrous dark gray bird apparently stuck in the mud on a riverbank with its wings flailing violently. The witnesses said that the bird had a least a 20-foot wingspan.

Source: www.thecryptocrew.com

Summer 1995 Stanley, Kansas- Giant Eagle

BALD EAGLE AT YELLOWSTONE NAITONAL PARK, WY BY LOADMASTER (DAVID R. TRIBBLE)THIS IMAGE WAS MADE BY LOADMASTER (DAVID R. TRIBBLE)EMAIL THE AUTHOR: DAVID R. TRIBBLEALSO SEE MY PERSONAL GALLERY AT GOOGLE PHOTOS - OWN WORK, CC BY-SA 4.0, HTTPS://COMMONS.WIKIMEDIA.ORG/W/INDEX.PHP?CURID=53834322

Richard Frieburg lived near 161st and Old Metcalf Road in Stanley Kansas in a new house at the time of this event, which he related to me in August of 2020 (see his other sighting August 4, 2020). One evening he was outside washing his car in the driveway when he suddenly had the thought to look up. He heard no sound but looked anyway. There, on the peak of the ridge of the roof was a huge eagle which he estimated at eight feet in height. When the eagle saw Richard, it took off and flew away. Richard just stood there and felt no fear or concern about it. Upon looking back at this incident, he finds this behavior rather strange. He did not recall having any missing time.

The largest eagle in the U.S. is the Bald Eagle which has a body length of 28–40 inches and the typical wingspan is between 5 ft 11 in and 7 ft 7 in. Since there is no such thing as an 8' tall eagle, we must rule that out as a possibility. Large eagles and owls are considered to be screen memories by many ufologists since many UFO experiences have reported seeing such creatures at the time of their sighting or abduction.

1996 St. Louis, Missouri- Pterodactyl Observed by Three Men

I received this report in 1996 and filed it away with my high-strangeness cases. A man and his brother and cousin were fishing in the Loutre River west of St. Louis, Missouri when they looked up to see a huge dark brown pterodactyl-like creature fly 300' overhead from east to west. They all dropped their fishing poles and stared, dumbfounded, at the creature, which then flew out of sight towards the bluffs. The witness said that the creature was gigantic, and it never flapped its wings but seemed to be gliding on air currents. They could not identify it as anything other than a pterodactyl due to the appearance of the beak, wings, and feet.

1999 Sedalia, Missouri- MothMan Sighting

From the National UFO Reporting Center: A man and wife saw a huge creature with wings flying slowly over the city of Sedalia. The creature was illuminated by the city lights. No further information is available. Sedalia is a town located 30 miles south of the Missouri River and 50 miles southeast of Kansas City.

2000 - 2021 SIGHTINGS

2000 Kansas City, MO - Humanoid with wings

Two women were getting out of their car at their home near Swope Park in Kansas City, Missouri when they saw a gigantic black humanoid bird with a human face and a beak. They were so stunned that they ran inside the house, dropping their groceries on the way. This home is the site of three sightings of strange, winged creatures, and very close to Swope Park which is known for strange occurrences.

2000 Independence, MO – Two Pterodactyls Riding on Air Currents

Frank L. from Northeast Independence, Missouri was getting groceries out of his car one evening when he suddenly had an inexplicable feeling to look up, and when he did so he saw two gigantic pterodactyl-like dark brown creatures flying high overhead. The pair of creatures floated on the currents and never flapped their wings during the duration of the sighting which lasted for at least 30 minutes.

Frank was too stunned to get his cell phone out and take a photo, but he said that there is no mistaking what he saw and that the winged creatures could not have been anything other than Pterodactyls.

February 19, 2001 Sedalia, Missouri – Giant Winged Creature

From the National UFO Reporting Center- a witness reported the following:

"Spotted an entity moving rapidly. I was parked at a rest area just northeast of town on the Katy Trail watching an airplane touch and go (I think it was a pilot in training with the trainer). I had been sitting there for an hour or so just relaxing and listening to geese, owls, cows and watching the airplane. I had rolled my window down and laid in the car sideways with my head resting partially out the window so I could see the sky. I saw a polar satellite go by at one point. I also saw a meteor which appeared to come from the constellation Cassiopeia moving towards the west. I was thinking about leaving when I spotted a dark object moving from the southeast to the northwest rapidly. It did not appear to have any lights and made no sound. It was in sight and out in less than 2 seconds. Needless to say, I had a hard time getting a fix on it. It was about the width of my pinky finger one inch from my eye. This object appeared to be less than 1000 feet high. Here is what spooks me... it looked like it was alive. It looked more like an entity than a craft which is why it changed shape."

Could this have been a giant bird rather than a plane?

2003 Toronto, Ohio – Mothman

Three friends were hiking on forest trails as the sun began to set. They heard a noise behind them and turned to see a terrifying creature. According to one of the hikers the creature was gray in color and had eyes lie the devil. The being then few up and swooped down ad the three hikers, nearly knocking them over. They ran down the trail to get away, but it followed them while making a horrible screeching sound. They finally jumped into a creek and the creature then disappeared into the woods.
Source: exemplore.com

Note: Many sightings of such creatures occur at dusk.

July 15, 2004 St. Louis, Missouri- Pterodactyl

As reported by Jonathan David Witcomb in his book and blog "Live Pterosaurs in America," "A man and his grandmother saw a large apparently smooth-skinned creature flying about a hundred feet above an Arby's restaurant in St. Louis. The witness said that the creature had the diamond-shaped tail end, and its wings were at least 20 feet wide."

2005 Rossendale Valley Lancashire England - Fairies Caught on Camera

John Hyatt, a Manchester Metropolitan University lecturer and Director of Manchester Institute for Research and Innovation in Art and Design captured fairies on film while photographing the landscape. These images have been spread widely over the internet and can be found easily. The pictures have been examined by professionals who say that the photos have not been altered. Hyatt went on to take photos of what he calls "Spirits of the Air," which look like ghosts. In one such photo, the image looks like Lady Macbeth. Hyatt says that it is his job as an artist to open people's eyes to the wonders through which they walk every day.

August 2005 Kansas City, MO- Huge Bat-Like Bird Jumped on Trees

As reported by Ken Gerhard in his book "Encounters With Flying Humanoids," Investigator Lon Strickler reported that two sisters in the Swope Park area of Kansas City heard a loud noise and looked up to see a gigantic bird with a 12' -18' wingspan hopping from tree branch to tree branch. It had leathery skin and batlike wings, a curved beak, with dark eyes. The two women ran inside the house and one of them fainted.

Kansas City's Winged Demon
From Ken Gerhard
"A seemingly relevant case that I would like to include is one that arrived on my doorstep via a paranormal investigator from

Kansas City, Missouri who didn't deem herself qualified to draw any conclusions. It involves two sisters who both encountered something that, they describe as a kind of demon bird. There are apparently at least four sightings by multiple witnesses and curiously, the primary contact is a woman named Angel who, confided in me that she happens to be a very devout Christian. I interviewed her and her sister at length over the phone and gleaned the following details.

Their first encounter occurred just two days after the great tragedy of Hurricane Katrina during August of 2005. Angel and her sister live in an area of Kansas City called Swope Park, which lies near the city zoo. As Angel's sister and a male friend were standing out in the yard talking one day, they apparently heard a very loud noise. Looking up, they noticed what they at first took to be a giant bird, hopping awkwardly from tree branch to tree branch. As it drew closer, they were able to discern that the 'bird' was black in color and truly gigantic in stature, at least the height of a man and displaying a wingspan between twelve and eighteen feet across. The creature possessed bat like wings and leathery skin, in addition to a long, curved beak and dark, deep set eyes. Angel's sister also told me that the apparition possessed a huge head with a face like an 'ugly person.' Whatever the thing was, it seemed to be attempting to become airborne, though quite unsuccessfully. The observers could not help but notice that the entity appeared 'wobbly' and unsure in its movements. When the incomprehensible beast eventually landed on their property, the pair frantically scrambled into the house where, Angel's sister collapsed on the floor in a state of shock. The next thing they knew, there was a loud thud on the side of the house as if, the monster had slammed into the side of the wall.

For years, the family tried to forget the terrifying events of that day until April of 2011, when the creature returned. This time it was in the evening and once again, the monster's appearance was preceded by a strange sound, likened to a dog being choked. The demon bird was then observed cavorting among the streetlights. An inquiry by the family led to word of another incident that involved one of their cousins who, claimed that he also saw the thing after leaving their house one night. He had spotted the animal when it hopped behind the taillights of his car and observed that it had a fat, round body, and a wingspan 20' to 25' wide. Another neighborhood couple stated they had seen the demon bird on Eastwood Street back in 2000. They swore that it had a humanlike face, but with a beak that had holes (nostrils?) in its side. During my extensive interview, I definitely got the sense that both Angel and her sister were emotionally scarred by their encounters and in dire need of answers. I could provide none."

Note that there are several reports which include a similar fact: some of the winged creatures, unlike the common birds we know of today, seem to have great difficulty moving around on the ground or getting airborne.

August 1, 2007 Minneapolis, Minnesota Mothman Spotted One Month Prior to the Bridge Collapse
Witness report seeing a Mothman one month before the Minneapolis Bridge collapse at the Interstate 35 bridge. Thirteen people were killed and more than 145 were injured.

The bridge crossed over the Mississippi River and carried 140,000 vehicles daily. It was built in 1967, which coincidentally was the same year of the Silver bridge collapse in Point Pleasant West Virginia.

The Minneapolis bridge is a part of I-35 and the Silver Bridge in Point Pleasant was a part of U.S. Hwy 35. Note that many of the Mothman sightings have occurred in Kansas City, where I-35 passes through from north to south. Could there be a connection?

People had been reported seeing a large bird-like creature for several weeks before the incident, but officials shrugged it off as a heron or crane and did not investigate.

On August 10, 2007 George Noory, host of Coast-to-Coast AM took calls from several individuals who reported seeing many strange creatures prior to the bridge collapse. One caller reported seeing a Mothman type creature with a huge wingspan on Wednesday, June 27, 2007, over a month before the bridge collapse, while driving outside of Stewartville, Minnesota. As far as I know, no other sightings have occurred in this area since and we may never know if the Mothman had something to do with it, or perhaps was just on hand to view the event.

Pterodactyl-like Birds, SW Missouri Unknown Date
The following report is from "A Menagerie of Mysterious Beasts" by Ken Gerhard with permission:

Missouri Mystery

I must also mention the American heartland and specifically the state of Missouri as a place where massive and mysterious flying things have been encountered. For example, there is the remarkable experience of a woman named Georgia Brasher who grew up on a 700-acre farm in the far southeast corner of the state. Georgia wrote me a letter in order to enlighten me about a chilling experience she had at age twelve or thirteen while, riding her dirt bike near a big ditch on the edge of the property –

"As I got closer to the tree line, I saw two huge black objects on two of the trees. One was bigger than the other one. My mind did not register that these two huge objects were birds. I thought something must have come off a plane and landed in the trees. The reason I thought that was the way the tree was bowed over. It was heavy and weighing down the tree… The closer I got it looked less and less like airplane parts… All of a sudden, the bigger one opened up its wings and hovered off the tree. I could not believe I was looking at giant birds. My immediate first thought was this is some prehistoric stuff. They did not look like they had feathers. It looked like skin when the wings opened, and the span of the wings was enough to scare me… They never took their eyes off me. I was not close enough to see detail in the face, but they had exceedingly long, bony looking shapes. I felt like I was being sized up and I knew that the size of those birds could have done me real harm. I got back on my bike and bugged out. I was looking back the whole time expecting to see them swooping down on me, but they stayed on the tree… When I got back, I was ranting like a crazy person, begging my mom to drive out there and take a look. She would have none of it… I didn't take my

eyes of the skies for an awfully long time."

2010 Grandview, Missouri - Gigantic Winged Creature Spooks Neighborhood for Weeks, then Moves to Lee's Summit

At 2:00 am one day in July of 2010 as Don S. arrived home from work and pulled into his driveway, he saw a huge 6' tall bird-like creature sitting on top of his roof. Too afraid to exit the vehicle, Don sat in the car and watched the creature for a few minutes, as it stared back at him with big eyes. The beast was a brownish color, had a chest area like a human, and two human-like legs, however, the legs were bent backwards at the "knees." It had a fluffy peak on top of its head and featherless wings like a Pterodactyl. The creature jumped downward, then flapped its wings just one time, silently, and became airborne, going in a straight vertical path upwards and into the night sky and out of sight in a seemingly impossible maneuver. The wingspan was an incredible 25'. Don thought that the creature reminded him of a creature from Jeepers Creatures.

Still shaken by what he had seen, the next day Don asked a neighbor if she had seen a strange big bird hanging around. She started laughing and called her husband out- they had both seen it several times and said that many of their neighbors had as well. Don spoke to at least 15 other persons who stated that they, too, saw this huge creature, and some called it a Mothman.

That is when things got even more bizarre. A single man moved into a house across the street from Don shortly after his first encounter with the giant bird-like creature. The new neighbor was rarely seen and did not talk to anyone. The man drove a white Econoline van with Maryland Plates on it. Don spotted the van on numerous occasions as he drove to or from work or around town, and it seemed to be following him. On several occasions, a message would appear on his cell phone which read: "FBI Surveillance," and he sometime heard a conversation through his car radio between two people whispering. Don cannot explain how these things happened. After five months, the man moved away.

During the two-year period that this creature seemed to hang out in the area, the children in the neighborhood spoke about seeing a strange, winged creature perched on the light poles during the day as they got off the bus in the afternoon or waited for the bus in the morning. They even gave it a nickname: "The Old Man." Oddly, the adults only saw the Mothman at or around 2:00 am in the morning.

Don continued to see the creature, as well as UFOs for two years, until he moved to Lee's Summit, Missouri. That is when things got even more strange. Don could inexplicably feel the creature's presence at times but did not see it. He also saw the white van following him a few more times.

One of Don's sons was driving home from work from Grandview to Lee's Summit at 2:00 am when he spotted a large, winged creature flying overhead. It opened it wings and displayed a completely black underside, but he could see no features in the dark.

Don had another encounter on Jackson Street in Lee's summit. While walking he felt something large pass right over his head. He could feel the whoosh of wind as it flew close to him, but he could see nothing. The creature, surprisingly, was now apparently invisible.

One evening in 2012 Don and his son were driving near Lee's Summit Road & 40 Highway in Independence, Missouri. They spotted the same creature and this time it followed them. Once again, it was 2:00 am in the morning, and the men were headed home from work. The creature then seemed to disappear into a portal and was no longer visible. When they arrived home, the two men saw two exceptionally large fish (carp?) which were still alive and flopping around on the ground by his mailbox. They wondered if the gigantic bird left them there.

Two weeks later, Don crashed his son's car. Two weeks after that, Don's son crashed Don's car. Both were severely injured and had life-threatening injuries but survived. For this reason, Don believes that the winged creatures are harbingers of bad things and that the car wrecks were caused by the winged creature. He also ties the two fish to the two car accidents.

Then large orbs started to appear at 10:00 pm on a nightly basis. These balls of light were either purple, blue, or yellow. The orbs appeared in the same location between Lee's summit and Independence in the northwest sky, at approximately 20,000 feet, for 1 ½ hours each evening for two weeks. They would then just blink out. After that time, the orbs only appeared occasionally. Don believes that the objects were UFOs and not a star, planet, plane, or helicopter. *Editor's note:* Multiple UFO sightings were reported in the same area at this time by independent witnesses.

One evening Don had a dream where he saw an orb, then it opened up, and something like the creature came out and tried to put its long pointy fingers into Don's forehead. When Don woke up in the morning, he had a bad headache. Since that experience he has had multiple premonitions of future events that came true, usually within two weeks. However, there is one vision he has had since 2012 that has not yet happened, and that is a disastrous earthquake in Missouri that levels buildings in major cities. Oddly, I have been having the same visions for the past 10 years. We both hope that we are wrong.

2011 Kansas City, MO - Huge Bat-Like Bird
Near Swope Park (continued from the report from the year 2000) The same creature returned in the evening and made a sound like a dog being choked, then cavorted among the streetlights.

2011 Kansas City, MO - Huge Bat-Like Bird
Near Swope Park. (continued from the above report) A cousin saw the same creature in his rear-view mirror when he left their house one evening. It hopped behind the taillights of his car. It had gigantic wings and a fat body.

May 14, 2011 Independence, MO - Huge Strange Bird/Pterodactyl with Puff on the Tail
Rachel T. and her two boys aged ten and eight were getting out of their car at 3:30 pm when they saw a gigantic dark brown bird with no feathers and a long tail with a puff at the end of the tail fly directly over their house in east

Independence, Missouri. The bird was approximately 100-200 feet above them. Her 10-year-old first pointed it out and said, "Look Mom!"

The witness said that the flying creature was the size of a small plane with at least a 25-to-30-foot wingspan. Rachel said that the creature made no sound and never flapped its wings. This site is located two miles south of the Missouri River and is in a wooded area.

May 22, 2011 Joplin, Missouri- The Butterfly People

A powerful EF5 rated multiple-vortex mile-wide tornado struck Joplin Missouri on the evening of Sunday May 22, 2011 causing massive destruction and the death of 158 people. It was the deadliest tornado to strike the U.S. since 1947. I was watching the news from my home in Independence as the tornado approached Joplin and sensed that something was not right about it. It was a heavy, dark feeling that is difficult to describe, but it felt as though the tornado was not natural or normal at all and I told my family that this was going to be awfully bad. And it was.

I spoke with other people who had the same feeling. While watching the news live, and later videos, I saw several UFOs moving about the tornado, and others saw the same thing.

A number of persons - mostly children- witnessed what they called Butterfly People during the tornado. The witnesses were many, and they and their parents believe that these creatures saved their lives since they were in situations that were not survivable, yet while everything around them was destroyed some of the people remained untouched. Some reports were of tiny fairy like people with wings, some saw white lights, but others reported seeing human or larger than human sized winged people. In nearly all cases the children called them "Butterfly People."

According to the Joplin Globe newspaper, in one instance Lage Grigsby,14, and Mason Lillard,10, encountered two people with wings. One had brown hair, the other had blond hair. The cousins were calmed by these creatures and even though both had severe injuries they both survived. Lillard said that as she lay in the wreckage of her grandparent's vehicle which had been tossed more than 300 feet across a parking lot, a hand touched her shoulder. She thought it was her cousin, Mason, but when she turned to look it was two Butterfly People.

In another instance a mother ran for shelter while holding her young daughter. They were knocked to the ground by the wind and the mother saw the tornado lift a car in the air. It was headed right for them. The mother crouched over her daughter, braced for impact, but nothing happened. The child said, "Didn't you see the butterfly people?" then told her mother that she saw butterfly people carrying people through the sky.

Joplin Tornado Source: Wikimedia Commons

In all such cases the children called the creatures they saw Butterfly People. Why did only children see these creatures? It may be because young children are usually much more aware of higher-dimensional beings than adults.

The Butterfly People made such an impact on the town that a mural called "The Butterfly Effect" depicting the creatures was painted on the Dixit Printing building at the intersection of Main and 15th Streets in Joplin. Dave Lowenstein, the lead muralist, included artwork from local children depicting people with wings and butterflies. I visited the mural in 2018 and there was an eerie feeling about that site that I cannot explain.

Considering that UFOs and strange lights were observed along with the tornado, is it possible that the creatures who saved some of the people were actually winged aliens? Whatever they were, the Butterfly People helped people during a horrible disaster and saved lives. If they were Mothmen, they certainly were on site for the disaster as they have been in the past, but contrary to their reputation saved many people from certain death.

Sources: The Joplin Globe Newspaper; https://www.stltoday.com/news/local/metro/cca48b1a-282b-587d-902b-cd5f09ca8516.html; https://mysteriousuniverse.org/2017/01/the-butterfly-people-of-joplin-missouri/

2012 or 2013 Cedar Lake, Oklahoma- Thunderbird:
As reported to thecrytocrew.com in 2019, a group of men were riding four-wheelers near the Canadian river when they saw something they could not explain:

"I heard a swoosh sound and looked to my right, which is a somewhat open field with a tree line running in the same direction, at the path but probably 300 feet wide. There was a giant black bird that scared the living s#$! out of me. I was afraid to keep going forward, seriously, because I thought it my pluck me off the 4-wheeler. It was flying just a little above the tops of the trees, but was not over them, because it was more in the field… Seriously each wing was about the length of a Honda Civic and the body of it was every bit the size of me and I am 6 foot and weigh 200 lbs.…

It was the craziest, scariest thing I have ever witnessed. I never knew anything like that existed."

Summer 2012 Ireland – Fairies Steal Wedding Ring

I got an email from a woman in Ireland who needed assistance finding a wedding ring that she left on the kitchen windowsill while she was doing the dishes. The ring disappeared and she could not find it. She asked me to remote view the incident, which I did. She hadn't mentioned anything other than the facts, so I was shocked when I saw three fairies come into her house through the open back door and go to the sill and take the ring. I told her what I had seen, and she was surprised as well. The door had indeed been open. I told her to leave some sweets and milk out for the fairies before she went to bed. The next morning the ring was back on the sill!

September 2012 Independence, MO- Unknown Giant Bird:

This is my own sighting. I was standing in the front yard of my office along with my daughter and grandson, who both work for me. We noticed a very large black bird with a large beak fly at approximately 800 to 1,000 feet overhead. We were dumbstruck at the sight. My grandson, Phillip, got three photos of the bird and tried to identify it comparing it to known birds by researching it online but could find nothing that matched. We estimated the wingspan to be at least 12' – 15' in width. This was obviously a bird of some type, but we cannot identify it. See the photo following:

GIANT BIRD IN INDEPENDENCE PHOTO: PHILLIP THOMAS

Fall 2012 Lake Jacomo, Missouri- Humanoid with wings
Sandra S. and her husband Brian were driving near Lake Jacomo in the evening when they saw a black or dark color 7'-8' tall humanoid creature standing 25' feet from their vehicle. It appeared to have bat-like folded wings and a human-type male body and glowing red eyes. The two were shocked and scared, and sped up to get away from the creature. The creature remained standing as they drove away. Lake Jacomo is located 20 miles east of Kansas City, Missouri where multiple sightings of UFOs, balls of light, and winged creatures have been spotted.

2013 Colorado Springs, Colorado Gigantic Bird
According to Mark Turner's Mysteriousworld.blogspot.com three separate witnesses reported seeing gigantic birds that made shrieking sounds. In one case, the witness experienced a percussion type of pressure in both ears followed by a loud 'flap" sound from the creature's wings as her ears depressurized. Could this have been a real Thunderbird?

2014 Lee's Summit, Missouri - Mothman
Tony Degn served in the military and is an ex-police officer. He is trained to be observant. He walked outside his home in Lee's Summit, Missouri one afternoon when he noticed an unmarked black helicopter flying nearby and thinking this was odd, took some video of it. It is unusual to see low-flying helicopters in this neighborhood and this is the reason he started filming it.

Shortly after the black helicopter event, Tony was walking in the same neighborhood off Woods Chapel Road and 291 Highway when he noticed a single low-level small cloud measuring 30' feet in width at approximately 300 feet in altitude in the sky. There were no other clouds in the sky, and he thought that was odd.

With his trained perception on high alert, Tony stopped to look at this unusual cloud when a humanoid with large wings emerged from the base of the cloud, flapping its wings while staring directly at Tony. The creature, who was all black in color with bat-like wings and a 15-foot or larger wingspan, stared menacingly at Tony, then re-entered the cloud and did not re-emerge. Tony was so perplexed at what he saw that he did not even think about attempting to take a photo with his cell phone. This type of reaction is quite common among witnesses to bizarre events.

Tony later had other sightings of what he believes are the same creature.

Summer, 2014 – Colfax, Iowa

According to Johnathan David Whitcomb, a minister was driving home from work one day when he saw a large bat/bird with a lizard line body and tail trying to fly. It was having difficulty getting airborne. The man said it had a dragon-line face, was a dark gray or black color, and was six to eight feet in length with a 12-foot wingspan.

Summer, 2014 – Colfax, Iowa

(Continued from the above minister's report) A man was driving at dusk when his daughter and a friend began yelling at him to turn around. They spotted a winged dinosaur that was attempting to get airborne. The man did not believe what they said and kept driving.

June 28, 2015 Eminence, Missouri - Bat-like creature

Eminence is a city in Shannon County, Missouri. Eminence is located in the center of the Ozark National Scenic Riverways, Missouri's largest national park and the nation's first protected river system. A man and his wife and 14-year-old child were walking on a trail in the park when they heard the rustling of animals running through the woods. They looked in the direction of the sounds and saw a man-sized creature sitting on a large tree branch approximately 50 feet away from them. The creature had a humanoid shaped body and dark brown body and wings, which looked like gigantic bat wings. The creature appeared to be staring directly at them. The family was so startled that they backed down the trail, keeping watch behind them as they very quickly walked and ran back to their car. The witnesses said they will never go into the woods again.

June 2015, Shannon County Missouri – Batsquatch
Dr. Emmet Reary, a cryptid and UFO investigator, reports that in June of 2015 one family came face to face with the Batsquatch during a weekend getaway to Shannon County. Driving down the gravel road to the Blue Spring Trailhead the family's mother, "Sara," suddenly shouted for her husband to stop the car. Sara reported this incident to MUFON. According to Reary:

"I interviewed her twice and my colleague Gary Hart interviewed her as well. Then he and I got together and compared notes. This technique has been valuable for detailed and complicated cases through the years for field investigators."

According to Reary's report, "She was staring at a winged entity seven to eight feet tall with leathery wings from its shoulders to the ground, black in color with pointed ears on a triangular head. Its yellow eyes stared at her making her feel extremely uncomfortable."

Dr. Reary and a colleague also did a search of the area for physical evidence, but nothing was found. "We searched the area, checked some of the local caves and we didn't find any footprints or anything like that," Dr. Reary says. "It makes you wonder about the nature of these entities, where do these entities come from, are they inter-dimensional, or what?"

This is the first time that this author has heard the term "Batsquatch," but given the appearance of the entity one can imagine why it was used.

Blue Spring is 14 miles east of Eminence off of State Route 106. It is on the Current River and is one of the most beautiful sites in the state. The spring is 310 feet deep and one of the deepest in the U.S.

March 2016, Boonville, MO. Thunderbird:
A man named Mike called in to Coast-to-Coast AM on April 1, 2016 to report that while he and his son were on a fishing trip, they heard something like running horses and felt strong vibrations through their feet. The man did not see anything, however, his son thought he saw a gigantic bird with a 25-foot wingspan. Mike said, "I now know why they call them a thunderbird because of the noise and the feeling they cause in the ground." To listen to the show, visit this link to the Coast-to-Coast AM website:
http://www.coasttocoastam.com/article/return-of-the-3-headed-person/

November 2016: West Alton, Missouri - Mothman

A woman and her son saw a strange giant bird/man-creature with legs on a riverbank. The creature moved very awkwardly with its folded wings on the ground, then dive into the river as if it were catching fish. The woman and her son were so shocked by what they had seen that they left the area immediately.

May 23, 2017 Loveland, Colorado – Gigantic Winged Creature
MUFON Case #107409

Thank you to Colorado MUFON State Director Katie Griboski for sharing this event. At 12:45 AM on May 23, 2017 a man and his son saw a huge, winged creature as they drove on a highway. Here is the witness' statement (with permission from the witness):

"Started off as a huge black bird far off in the sky shaped like a crow but bigger than any craft, it approached fast and changed shape to an upright black see-through curtain type of being."

"I really do regret not reaching out immediately after the event happen. It left us feeling so weird did not know when to contact anybody. The reason I am reaching out now is we are hoping that somebody else other than my son and myself have seen something similar because we haven't been able to find much about it on the Internet other than it was a spirit bird or I guess Thunderbird. So, here is what happened in a nutshell. My Son and I were traveling home from Greeley on I-25 southbound going through Loveland at approximately 12:30- 1 o'clock in the morning and far off in the distance we notice a Black bird in the sky traveling our direction. The size was maybe as big as a cloud but dark enough to be able to see its shape and features. It approached very fast, there was also a semi-truck next to me, so I was trying to be cautious not to hit him, but I know that gentleman had the same experience we did that night and somehow kind of feel connected to him. The entity passed by the middle median and at that point I braced somewhat for impact and got ready to avoid by swerving. It passed by my window but still trying to watch the road could only see a large black drape style thing right next to me passing by but can say as it approached it changed shape to a tall upright figure. My Son, who was 19 at the time saw much more detail than I. This is his testimony:

"I saw a big black bird like thing at first then as it passed by the window, I saw the long drapes like my dad did then I saw it change into a long dark black figure taller than a house with dark eye holes for eyes it looked back at me as if the entity was trying to connect with me."

After it passed, we both remained absolutely silent until our exit, while sitting at the off ramp I gathered my bearings enough to talk and before explaining what I saw quickly asked my son to describe in detail what he saw. I made sure to quickly look at the clock because for some reason it felt like we should have been missing some time, but the clock was still correct. After he gave his description as to what happened it confirmed I wasn't crazy, and we both shared the same exact experience. We will never forget this event and feel that very few have ever seen the same entity." (Edited for clarity).

November 2017: Salem, MO: Three giant bird sightings

Investigator Emmett Reary from Salem, Missouri said that he has had a number of giant bird sightings reported to him from the Dent County area. Two witnesses saw birds with the wingspan of the house it flew over. Several other witnesses east of Salem, MO saw a giant black bird that swept down on them while they were having a barbecue at dusk.

January 12, 2018: Independence, MO: Supernatural Giant Winged Entity

As M.C. left work in the evening from her office near 23rd Street and Pearl Street she heard sounds like something was flapping and running on the ground towards her from behind. She thought the flapping sound was a plastic tarp that was tied over a crate, but she thought the running sound was someone chasing her. She did not look back - but ran to the car door and jumped into the driver's seat, then quickly closed and locked the door.

At that point, the woman noticed that there was something black in her rear-view mirror, then her entire Dodge Charger became enveloped in very large feathered black wings from the back around the side windows, and finally covering the front windshield. She was petrified. She could not see anything through the windows except blackness and could not see the body of whatever it was. Then the wings appeared to dissipate, and the creature was no longer there.

A check of the security camera at the business showed nothing on camera during the time of the event, other than the woman running to her car and getting inside. Since there was no wind that night, the flapping sound could not have been the plastic tarp. What could have made sounds and be visible to this woman, and not show up on a security camera? Could this have been an inter-dimensional creature? The oddest part about this report is the dissipating wings and disappearance of the creature rather than the witness seeing it fly away.

January 13, 2018: Independence, MO. Thunderbird and Three Smaller Creatures.

This is my own sighting. If I had not seen it myself, I would not believe it. I was outside on my side deck in the late evening when movement caught my eye - I saw a gigantic set of dark grey wings approximately 25-30 feet in width and stretched out in the trees approximately 150 feet away. As my eyes adjusted to the dark, I could see a central figure standing on a big branch and the edges of the outstretched wings were rounded. The only light in the area came from a streetlight 300' away. Something made a noise in a tree near me, and I looked in that direction to see three smaller (24" tall) gray birds with closed jointed wings in the tree, then looked back at the large creature but it was no longer there. There had been no noise of a creature moving so I was perplexed about this. I looked back at the smaller tree, but the three smaller creatures were also gone. How any of these creatures were able to disappear without making a sound is beyond my comprehension.

I noticed that there was no wind, and it was eerily dead calm, as it was the night before. I could not tell if the creature left, or if it may have brought its wings down and was still sitting in the trees, but I decided not to stay to find out. This was just too weird for me, so I went inside the house and locked the door. The next night, prior to a storm, I heard strange thunder that went on and on without stopping for about two minutes. I was on the phone with a friend who lives 20 miles away and she heard it as well. She said she had just heard about a Thunderbird sighting and wondered if that could have been related to the weird thunder. This was before I told her about what I had seen the night before! Was this just a strange coincidence?

THREE WINGED CREATURES IN TREE DRAWING BY KRISTINA MCPHEETERS

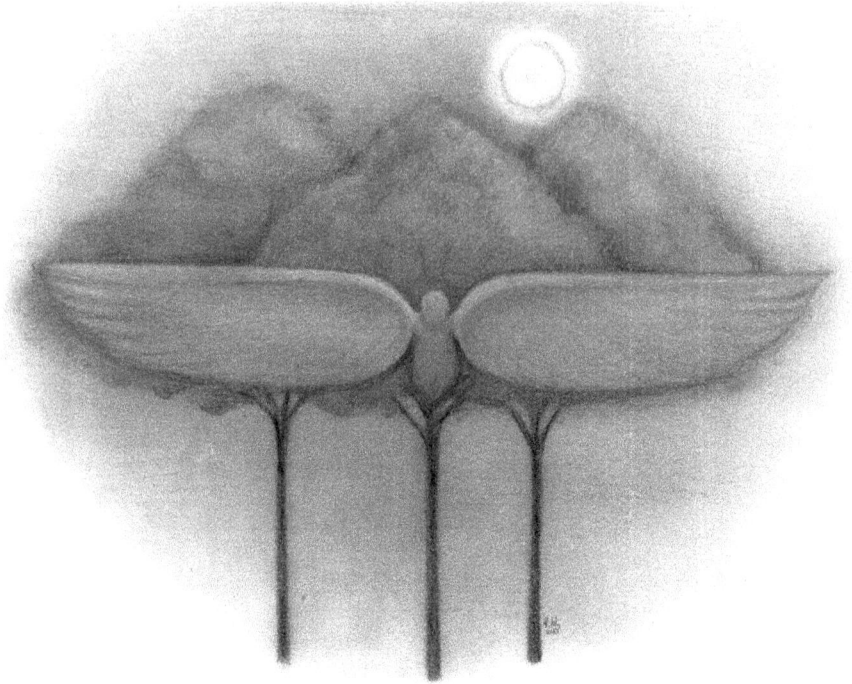

GIGANTIC WINGED CREATURE IN TREES AS SEEN BY MARGIE KAY
DRAWING: KRISTINA MCPHEETERS

This creature was so massive that I find it very difficult to believe that it was some type of conventional bird. Now the reader will understand why I am so interested in getting to the bottom of this strange phenomena.

Spring, 2018 Tonganoxie, Kansas: Pterodactyl Sighting

Hunter Leihy lived on a 147-acre farm near Tonganoxie with his grandparents at the time of this event. In the spring of 2018, he and his girlfriend were out sky watching when they noticed a strange yellow/white light in the sky. At first, they thought it was a planet, but then the object began to get larger as it moved closer and began to move very quickly from side-to-side and up and down. At that point, they realized that it was not a planet or even a drone due to its amazingly fast movements.

A few days later, the couple was driving in the field one evening to sky watch again and to see if the strange light would appear once more when Hunter noticed a very large bird flying above their vehicle. He pointed it out to his girlfriend, who also saw the creature which then swooped down over their Saturn Ion from the back to the front and almost touching the windshield. The couple thought the winged creature might even attack their car. They noticed that the size of the creature was five to six times the size of their vehicle. As it glided overhead the couple noticed that the creature had leathery egg-colored body and jointed wings which were pointed at the back, and that it had a long beak with a cone on the back of it. The two believe that what they saw was a pterodactyl.

The area is heavily wooded and has many creeks and springs. Hunter's grandparents and their neighbors have had many missing chickens, dogs, and cats in recent years which have never been found. If a large cryptid creature is living in the area it would certainly have plenty of food and water sources, and places to hide.

Summer, 2018: West of St. Louis, MO- Flying Creature with a Human Head

Michael R. was driving near his rural property 50 miles west of St. Louis when he saw a giant winged creature with a human head flying on the highway in front of him and his vehicle. The creature was being chased and attacked by drones. He watched this for approximately 15 minutes as he drove, then the creature veered off into the woods and out of sight.

This location is the site of many strange and unexplained events. Michael has had a number of sightings and I have obtained trace evidence from the area that cannot be explained. There is much UFO involvement, and we believe that this creature and others are alien related as we cannot find a mundane explanation. The fact that drones were chasing the creature is evidence that someone else knows about these beings, and the most likely candidate is our own military, who were likely piloting the drones.

November 2018: Blue Springs, MO: Huge Flying Creature

Samantha Belvior saw a gigantic black, winged creature swoop down over her car in Blue Springs, Missouri. Its wingspan was as big as the length of the car. In her email which is reprinted with permission the witness stated the following:

"Hi Margie Kay,

I live in Blue Springs and in Nov 2018 I saw a massive black winged creature swoop down out of a tree on Briarwood in Blue Springs, MO. It flew over my car for about 50 yards and disappeared. The wings were huge as they expanded well beyond the width of my car. I never saw a face, so I do not know if it was humanoid in appearance, but I have lived in Apple Valley for 20 years and that is the first time I witnessed anything like that.

This happened on two consecutive Thursday nights at the same location in Nov 2018, which I thought was odd, why Thursday?

I do not know if that helps with your records, but I have had that information for a while and had. I idea who to share it with. Hopefully, you find it helpful.

Kind regards,

Samantha Belvoir."

January 2019 Mothman Sighting in Washington State

The following report was submitted by MUFON Field Investigator Jeannette La Tulippe:

Part of the following has been paraphrased through an article written by James Gilliland at ECETI (Enlightened Contact with Extraterrestrial Intelligence) ranch in the state of Washington, a place where UFOs are seen with regularity. We can no longer afford to live in denial of unseen/unacknowledged negative influences. Our denial of such an acknowledgment allows them to roam and to cognitively influence persons freely, a carte blanche if you will. Has our fascination with the phenomena overruled our better judgment? Are we mesmerized with the storyline as opposed to what makes these events occur, displacing how important this really is? If a person has passed this does not make them enlightened, and just because a being is an Extraterrestrial, which may be perceived as a higher vibration/frequency, does not mean it is benevolent. Their frequency is simply its state of being as ours is to us. A saying in India states: The closer you get to God and/or the closer you get to Nirvana; demons will rear their ugly heads frequently into your life. To be enlightened means to have the knowledge of both the light and shadow sides of a thing. Without the knowledge of both, we cannot profess to be enlightened but, instead, we live a half- truth.

On January 06, 2019 case #97499 was submitted to MUFON. I was one of the two investigators interviewing the witness, who had served in the military and was currently working in a correctional facility, by which he had to have had a background check to qualify for his position.

Essentially, what he reported was a seven-foot tall or more, dark gray, leathery, large winged (like that of a bat), humanoid creature landing in front of his car, during good weather, at 4:10 a.m. The creature dropped down from above without wing flap. The witness could not see its eyes nor its face. Throughout the entirety of the sighting, it did not exhibit flight but rather it leapt.

By wrapping its wings about itself bracing as if for a plausible impact from the witness's car which was traveling at 85 mph, it exhibited intelligence. What may have been called a cryptid, its wings were one unit, meaning, it had no arms per se but only had what were seen exclusively as wings. Its chest was too small to have taken in sufficient breath for a sustained flight and its wings were in what is considered to be the wrong position to enable flight. In addition, there seemed to be no way for it to take in sustenance. The witness had no memory of seeing its feet. I have asked other people that have investigated these cases and they also do not recall seeing feet on these types of creatures.

What is it? what is that? I can't see any eyes. what is that? It has no face. What is that? What is it. I keep thinking. I only have 4 seconds to react. So I start easing my foot off the accelerator and turn the wheels slightly to the left to avoid hitting this thing.

CREATURE STANDING IN THE ROAD DRAWING BY THE WITNESS

The musculature in its legs was substantial which would account for the ability to drop down from a distance, especially by slowing itself with its wings absorbing the shock from a fall. The leg musculature indicated that the muscles looked as though they were often stretched. Prior to the sighting, two green orbs flew from in back of the witness's car truncating their trajectory only to go over the relatively small hills to his right. He heard no crash and believed they flew underground down into the mine shafts.

CREATURE AS CAR APPROACHED DRAWING BY THE WITNESS

At one part on the road, I experienced an odd feeling, as if time were slowing. Driving with his car in front of ours, the witness signaled us to pull over. He got out and explained that every time he had driven on that section of road there seemed to be a sensation that time slowed, despite the fact that his clock always read correctly.

Close by, there was a nuclear testing area that had been used in the past. I told him I experienced the same experience of slowing time; he felt happy that I concurred. There are texts that will verify that the frequencies generated by an atomic blast open a porthole, the porthole to the abyss, not a positive place by any means. The Necronomicon is one such book.

At this point it let out this loud annoying frequency like, that sounded like a bat in extream pain and distress. It was at this point it showed me pictures of animals crossing the street. I could see throw these animals. It was a rabit a bird and a dear. Then It said in a very clear voice. I am no different than these. I am no threat to you. All this at the same time. I was thinking about shooting this thing.

This happend 27th September 2018 0410 A.m.

My thoughts and its answuss almost Simultaneous. I was not the only car on that road that morning. why me?

2021

CREATURE FLYING DRAWING BY THE WITNESS

The following picture depicts the hills the green orbs flew over, the area of the witness's first sighting, and to the right of the wood fence posts is the Native American graveyard.

FIRST SIGHTING PHOTO BY JEANETTE LA TULLIPPE

We revisited the area where he had the first sighting. To see if I could pull up any additional information, I got out my ghost hunting equipment. It indicated that we were standing by a mile marker, my tool spoke about the witness's ethnicity, and told me that we were by a Native American burial ground. My tool was accurate in its description of the witness's ethnicity and the mile marker. I prodded the other investigator to ask the witness if we were by a burial ground. The witness pointed and said, "it's right there" and that if we wanted more information on it, we could go to the trading post in town.

The witness also said that at one time, while looking at the cryptid, it was as though he was looking through a spyglass with the exterior surface on the inside…meaning, that it seemed to jump further away in stages as if a poor editing job had been done with film. During this period, the creature was surrounded by some type of fog, a fog I call an electronic fog, one that has been present when there are issues with time e.g. The Hutchison effect, a cloud of electromagnetic fields whereupon all of the electronic systems and instruments malfunction. Spatial disorientation and time distortion starts until the object finally disappears without a trace. After having experienced this…it was gone.

At this point I would like to mention there are several similarities in Mothman, Dogman, and bigfoot cases. Putting the notion of a physical being aside for now…there is a likelihood of a thing masquerading as another. The Native Americans call them tricksters or Skinwalkers. Names outside of their culture may be liar and deceiver. Accomplished occult practitioners know that spirits need to be controlled because they are often liars and can be wild and destructive in their actions. To control them a talisman is used. Demons' extremities are not seen. The lower portions of their arms or legs seem to disappear. In this case and in others the feet of the Mothman were not seen.

Because of the other things I deal with, the archangel Raphael has appeared on occasion. His (male energy) appearance was similar to that of a being in a Mothman case I read about, in which the creature had large lumps on the back of its head which may have been eyes; it was seen in the dark which is why only the lumps were spoken of. To be exact, Raphael has six eyes on the back of his head. I should also mention that this is more or less an insectoid trait. My question is, is Raphael an insectoid (alien) being? Raphael is an archangel of the seraphim class of (alleged) angels. Some of us investigators are questioning if what we are investigating are not in fact the angels and demons/aliens of old capable of shapeshifting, amending their appearances, thereby keeping with the times.

Below you will see a picture of what looks like an antelope that is alive and is standing on its feet. In the picture you will also see what looks like a spirit either entering or leaving its body, a plausible possession or shapeshift in progress.

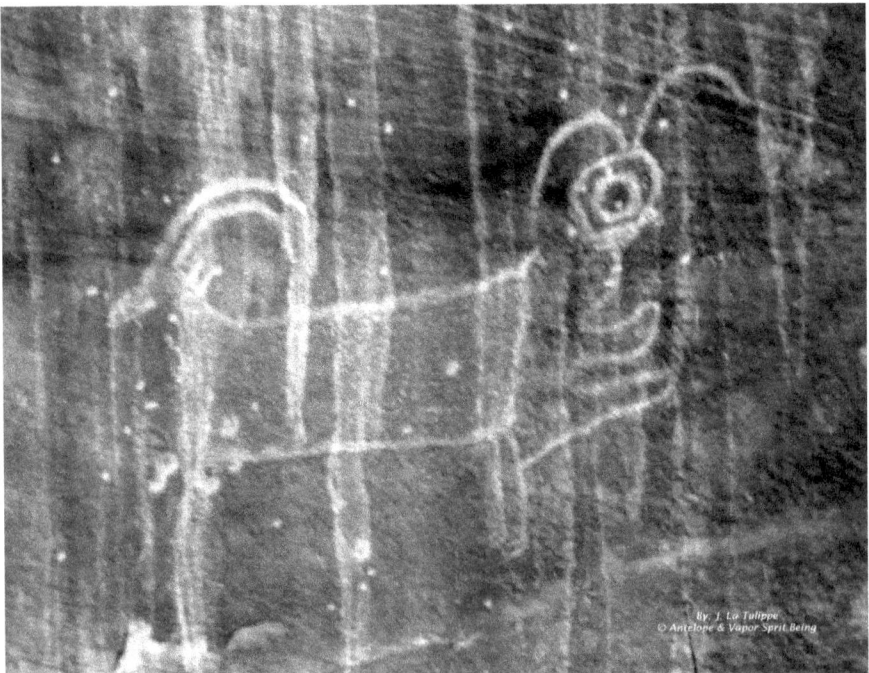

by J. La Tulippe
© Antelope & Vapor Sprit Being

February 22, 2019 Woodstock, Illinois - Large Furry Creature with Wings

According to the Singular Fortean Society, a man reported seeing an eight- to nine-foot-tall creature covered in dark fur with large wings. The man thought the creature may have been a bigfoot, however, it had membrane-like wings attached to the back and extending over the top of its head.

Investigators Tobias and Emily Wayland visited the Woodstock location and the site where a creature was spotted in November of 2020. They did not notice any radio interference either time, however, upon review of the footage they recorded during their initial visit, they did notice some very unusual audio interference on both cameras used during the investigation. The same interference was present upon review of the footage taken on their second trip, although it was much lighter. Strangely, the interference was only recorded in the immediate area of the sighting.

June 20, 2019 Sugar Creek, Missouri - Thunderbird?

Valerie Hoover was working on her garden when she heard an aircraft flying overhead and looked up to see not only as small-engine plane, but a huge black bird with a huge wingspan flying above the aircraft. Valerie said that the creature never flapped its wings or made a sound. She said that the creature was the size of the small-engine plane underneath it (approximate 36' wingspan). It was flying too high to see details.

This location is the site of multiple UFO sightings by not only Valerie, but investigators including myself and Missouri

MUFON Kansas City Section Director Jean Walker and others. This site is approximately ½ mile from the Missouri River and close to Kansas City and Independence. Coincidentally, on this date there were thousands of UFO sighting reports in the greater Kansas City area which were highly publicized and remain unexplained.

June 20, 2019 Blue Springs, Missouri - Huge Bird

Terre Tweedie heard birds screeching outside her house in Blue Springs, Missouri and thought a hawk must have been after the birds. She ran outside and was astonished to see a huge unidentifiable black bird with a 10'-12' wingspan flying away. She could not identify the bird even though she is an outdoorswoman. The other birds kept squawking for a couple of minutes after the large creature was out of sight.

On June 20, 2019, two unidentified objects hovered high over the city for six hours. The objects were captured on video and the story was covered on national news. Some people thought they were weather balloons, but the National Weather Service said they were not weather balloons. DARPA released a statement that they released high-tech high-altitude balloons on June 17, three days prior to this event. However, the analysis of the video shows multiple objects flying around the larger white objects at incredible speeds.

On the same day, Quest Paranormal Investigations received reports of unidentified objects that looked like rods in South Kansas City, Missouri, a report of red orbs in the Waldo area of Kansas City, Missouri, and an unexplained brilliant blue light in Independence. The entire K.C. area has been a UFO hot spot since 2011. Could these things be related as they seemed to be in Point Pleasant prior to the bridge collapse there?

June 2019 Joplin, MO – Giant Humanoid
As reported to Lon Strickler:

A woman driving from Springfield to Kansas City drove through Joplin sometime between 7:00 and 8:00 pm, before it was dark. There were thunderstorms before and after this sighting. The witness mentioned this because he suspects that the electric charges of the thunderstorm had something to do with his ability to see the creature.

The witness saw a huge, winged creature with a wingspan of 12 – 16 feet flying low just above her vehicle. The size enveloped the vehicle. She could see the wings and part of the torso through the sunroof and windshield before it flew toward trees. The witness did not hear any sound. Oddly, the witness was miles down the road when she suddenly realized that she was perfectly calm until that moment, and then suddenly in shock about what she had just seen. The witness mentioned that she is a UFO contactee, which may be significant.

More information: https://www.phantomsandmonsters.com/2020/05/huge-winged-humanoid-encountered-in.html

July 29, 2019 Independence, MO - Giant Bird
A man by the name of Alex W. was driving south on 291 Highway in Independence, Missouri at 5:00 pm when he noticed a very large bird circling high overhead. He pulled over to get a better look at the dark-colored creature because the

size was so amazing. Alex said that it was similar to the shape of an Eagle, but way too large to be a bird of that type and that it must have been something else. He was unable to get a photograph. Alex said that the skies were clear and sunny, and he got a good look at the creature. 291 Highway is located on the east side of Independence, Missouri and crosses the Missouri River. This location is not far from other events on 291 Highway.

July 30, 2019 Overland Park, KS (near KCMO)- Giant Bird

Dr. Jean Harty and her husband were driving on a busy street in Overland Park, Kansas when they were both shocked to see a huge black bird with a 15' - 18' wingspan fly down to the ground and into some brush. The couple could not identify the type of bird and were unable to stop due to traffic. At first, they thought it was a Condor, but realized that this creature was too large to be a Condor.

Condors are not in this region of the United Sates. Other people encountered a gigantic bird in the general area near this date (see the list of reports).

Summer, 2019- Kansas City MO- Giant Bird at the Downtown Airport

George Beal, a retired engineer, and his son were driving near the downtown airport near the Missouri River in Kansas City one afternoon when they noticed a large black bird with huge wings flying high overhead. Stunned at what they were seeing, the men stopped the car in a parking lot and got out to watch the creature floating on the air currents. They both thought it looked like a pterodactyl with a huge wingspan. It did not flap its wings, and slowly moved off towards Kansas to the west and eventually out of sight. It had no feathers and large beak. "I still can't believe what we saw," said Beal. "The thing had to be exactly what it looked like – a prehistoric bird." Beal and his son were so stunned by what they saw that they forgot to get video with their cell phones.

Summer, 2019- Wempletown, Illinois- MothMan/Batman

Two young men went out for a drive on the back roads near Wempletown, a small town north of Rockford. At approximately 10:00 pm as they slowed for a dip in the road the two noticed that the tall corn was shaking violently to the right of them. The corn suddenly parted open and a tall man walked out, spread its gigantic bat-like wings, and went airborne. The witness described the creature has having a shiny black tar sheen to it, so they called it the 'Flying Tar Man.' The two never told anyone about the incident for fear of ridicule. *Source: www.phantomsandmonsters.com.*

August 8, 2019: Chicago, Illinois Winged Humanoid at O-Hare International Airport

A pilot for a commercial airliner witnessed a large human with enormous wings and lowing read eyes perched on a rail. The pilot was a passenger in a shuttle bus at the time and watched as the creature swiveled its head and followed the shuttle as it passed approximately 15 feet away from where the creature was perched. The witness describes the creature as being all black, very skinny with bat-like wings and glowing red eyes.

Summer, 2019: Rolla, Missouri- Giant Bird-Shaped Metal Craft

Lowell Hill saw something he could not believe one sunny morning in August or September of 2019. From the witness's email:

"Well, what I saw was what looked like a big black bird. It was up pretty high, so it looked exceptionally large. It came out from my right side as I was facing southeast. I noticed it was moving fast, and the one thing I noticed was it was metal, and the wings was never flipped. It was not a bird, but a bird shaped craft. It flew east of Highway 65 heading ENE towards Rolla, Missouri. It was very large - about size of a private jet."

Mr. Hill stated that he heard no sound and that the craft flew in a straight-line trajectory then turned towards the east. He watched the craft for three to four minutes and was surprised that the object did not move its wings as a bird would.

There are a few people who have reported that what they saw looked more like a metallic craft than an actual bird. What the purpose could be for such a craft is as yet unknown. However, a recent Television documentary showed that there are some experimental aircraft being tested by the military that do have bird-like wings so perhaps this would explain some of the sightings.

October 2019 Cape Girardeau, Missouri – Black Humanoid with Large Wings and Legs

A woman, her fiancé and her two children were driving around town near the Southeast Missouri State Softball Field looking at Halloween decorations on houses when they came upon a tall black figure near the road. The head was rounded, and the wings were pulled in. The witnesses said that the wings looked like thin skin like bat wings. It reminded them of the creature from "Jeepers Creepers."

Source: https://www.singularfortean.com/news/2020/10/20/woman-reports-black-humanoid-figure-with-very-large-wings-and-legs-in-cape-girardeau-missouri

November 2019 Independence, Missouri - Hawk-shaped Drone?

I saw a hawk-like creature with a normal wingspan fly down over my car, and swoop close over my windshield and hood. I have never seen a hawk do that but would not have thought any more of it until I noticed a humming mechanical sound like a drone which was clearly coming for the flying object. Unfortunately, the faux bird flew quickly out of sight to my right, and I could not see it any longer. What possible motive could there be for a mechanical bird to dive-bomb a vehicle?

There are mechanical hawks. As an example, Clear Flight Solutions based in the Netherlands builds these robotic birds of prey designed to scare off unwanted birds from airports, farms, and downtown buildings. So, the technology exists, but why such a thing would be at my office in Independence or buzz my vehicle is a mystery.

December 12, 2019: Independence, Missouri
Mothman Intimidates Driver

While driving south on 291 Highway in Independence, MO Tony Degn saw a humanoid with gigantic bat-like wings fly towards his vehicle, stare at him, then fly away. Tony was headed home after work at KCI Airport at 3:00 a.m. when he noticed a gigantic bird in the distance. The creature headed in his direction and as it got closer Tony could see that it had a humanoid body with long legs and large bat-like wings that measured approximately 10 to 15 feet in width. He did not get a good look at the face or notice eyes. "I was focused on driving and on the wings and legs, which were quite amazing," said Tony.

Degn was driving south on Hwy 291 just past 24 Hwy as the creature swooped down to just 20' above his vehicle, then flew off to the east. "I noted that the wings did not operate as normal bird wings and that they reminded me of bat wings or Pterodactyl wings," said Tony. He thought the entire experience was very strange, especially since he saw a similar creature in Lee's Summit in July of 2014. His first thought when he noticed the creature on Thursday morning was "Oh no, it's back." He was unable to capture a photo of this creature.

December 2019: Trucker sees a seven-foot-tall humanoid with wings at O'Hare Airport

Manuel Navarett of UFO Clearinghouse took a report from a man who spotted a seven foot-tall "person with wings" just outside a fence by the parking lot at O'Hare International Airport in Chicago on November 26th, 2019. See the article here: https://alien-ufo-sightings.com/2019/12/trucker-reports-seven-foot-tall-person-with-wings-near-ohare-international-airport/

December 31, 2019 4:30 pm. Overland Park, Kansas- Huge Bird

Pat Delap saw an all-black winged bird flying overhead as she drove on College at 119th Street in Overland Park, Kansas. The bird was at treetop level and it was gliding. Pat said that it did not flap its wings and that it looked similar to a raven except much larger. She could not estimate the size, except to say that it was the largest bird she has ever seen.

December 31, 2019 7:50 pm Kansas City, Missouri- Giant Bird

As she was driving north on Wornall Road in Kansas City, Missouri at 7:50 pm Marlene Asbury saw a gigantic dark flying bird that looked similar to a pterodactyl. It had a huge beak, long legs with feet, and a wingspan that would have been as wide as the street. Marlene estimated the size to be the length of her minivan with a wingspan of approximately 20—25 feet. The creature flapped its wings several times, then Marlene lost sight of it as she drove.

Could this have been the same creature observed at 4:30 pm by Pat Delap? It seems possible since the two locations are just a few miles apart.

January 2020: East Jackson County, Missouri - FOUR Sightings at an Undisclosed lake area in Jackson County. Pterodactyl

R.H. and her sisters have seen a giant bird four times since 2016. The sisters live near a lake in Jackson County, Missouri and often do birdwatching together. The first time they spotted the creature it was sitting out on the ice and never moved. It looked like it had jointed wings and was supporting itself with the elbows of the wings. It looked like a prehistoric creature.

The next two times they saw it the creature was in flight, and it looked like it was the size of a small plane, but it flapped its wings like a Pterodactyl. The fourth time they observed the flying creature was in January of 2020 when the women were looking for eagles with binoculars and saw what they think is the same creature flying so high they could not see it with the naked eye. The women do not wish to disclose the exact location because they do not want any harm to come to the creature they saw. Both women, who are familiar with birds in the area, insist that the creature is a pterodactyl, and it cannot be anything else.

February 11, 2020 Sugar Creek, MO- Bird Morphs into a UFO

Valerie Hoover saw a hawk-like bird morph into a UFO and take off at incredible speed. Valerie was standing in her backyard in the afternoon when she noticed movement above and glanced up to see what she believed to be a large hawk. She watched it as it flew above her, not thinking much about it when it suddenly changed shape into a typical small disc-shaped silver UFO type craft, then instantly shot off and out of sight to the Southwest in less than a second. Valerie was stunned at what she saw and contacted me immediately. She said that the object made no sound.

This sighting is remarkable and is the only such report I have received in 40 years of doing paranormal and UFO investigations. I know Valerie personally, and can say that she is very observant and credible.

The fact that a flying creature changed shape into something entirely different could be evidence for an alien/UFO connection.

February 1, 2020 Lee's Summit, Missouri - Huge Bird
A woman in Lee's Summit, Missouri reported seeing an exceptionally large black bird clinging to the top portion of a light pole near her house. It was a foggy night, and she does not recall seeing details about the creature, only that it was unusually large and had wings that were folded and not extended. She also stated that there was a green electronic fog around the creature but nowhere else. She was frightened and went inside her house, but later thought about how strange it was and went back outside to see if the creature or the green fog was still there, but they were not.

March 16, 2020: Seneca, MO- Alien/bat/bird Creature Bites Man in the Face
Robert F. was camping on his land in a rural area near

Seneca, Missouri when he had the feeling that something was watching him. He started a campfire, then looked up from his seated position on a log and saw a huge transparent winged creature with humanoid legs and bat-like face and ears standing on the ground just ten feet away from him. The creature seemed to be shocked that the man was able to see it and approached him menacingly, then bit the man on the face, drawing blood before Robert could react. The creature then ran off and disappeared into the woods. The creature appeared again in September of 2020 at the same location but disappeared into the woods after hissing at the witness. It again showed itself in January of 2021, but this time it flew up into a tall tree without flapping its wings. Robert was dumbfounded at how the creature accomplished this feat without flapping its wings.

The witness believes that the creature may have been an alien rather than a bird/man. There have been many sightings of UFOs, Aliens, and Sasquatch at this location which I investigated with the assistance of other MUFON investigators. We have obtained physical evidence of damages to trees that could only occur from something like microwaves. This site is indeed one of the strangest in Missouri and I do not doubt the witness' statement to be true.

May 14, 2020 South Kansas City- Giant Bat

Rob L. lives in a rural area on 10 acres south of Kansas City. He was lying in bed and looking out his window at approximately 10:30 pm when he saw a huge black bat-like creature with bat-shaped wings and a 3' wingspan fly towards his window, then swoop away. The next night he saw a string of lights outside in the sky, which broke into multiple-colored lights. R.L. believes that the lights were UFOs and may have been related to the bat creature he saw.

Rob has a history of encounters with balls of light and UFOs. (Note: A check with MO Conservation indicates there are no bats that size in Missouri – the largest has a wingspan of 13.5")

May 14, 2020 South Kansas City, MO- Mothman

Lerone Pryor was visiting a friend on 81st Street near 71 Hwy in Kansas City. The two were barbecuing outdoors when Lerone noticed a strange shadow on the ground. He looked up and saw a man-sized humanoid creature with huge wings and a 3' long tail flying silently overhead and heading away from him. Lerone did not notice any features as the face and front were not visible. The creature was gliding and did not flap its wings or make a sound. It flew low at treetop level (approximately 60 feet or less from the ground) and out of sight. The witness stated that this area is known for UFO sightings and strange events.

This area is close to Swope Park, which is the site of many UFO, bigfoot, and Mothman sightings (see the last chapter for more information about the Swope Park area).

July 3, 2020, 8:60 pm Waldo area in South Kansas City, MO- Gigantic Bird

Mechanical designer Greg Graham and a friend spotted multiple bright white lights in the sky, then saw a gigantic bird the size of a Cessna airplane fly overhead from southwest to northeast.

From the witness' email:

"Wow. So glad I was out tonight! Was with a friend and we saw the only way is to describe it was a GIANT bird. About the size of a Cessna and about the same altitude. The moon was bright and illuminated it clearly. We could see the giant wings flap and it changed direction which just like a bird. I have ALWAYS thought the stories were bs. Wow! I cannot believe what we saw. We also saw around 20 immense bursts of light above Waldo. Moving the same direction south west to northeast. We googled pictures of star link and they were not a constant brightness and in a perfect straight line. They were sporadic and spaced out. We also saw a few satellites as reference, and the bursts acted nothing like them. We're both sitting here in amazement!!!"

Mr. Graham filed this report with MUFON, and I did the investigation. His statement to MUFON is as follows: "It was Thursday July 2nd at around 9:45 pm. My friend and I went outside in hopes to see the Starlink Satellites which were scheduled to fly over Kansas City. Our eyes were to the sky trying to find the train of satellites. Shortly after going outside, we noticed something flying high and focused on it. We thought it was a plane, but then we both saw it flapping. It flew directly over us at about the same height of a Cessna. We could clearly see the wings as it was a close to a full moon and there was no cloud coverage. We watched it fly overhead SW to NE, then before it disappeared past the tree line, it turned to go east, then swiftly changed direction and headed north. It changed direction just like a bird. After discussing with my friend, we realized just how big it was. To be the size it was at the height it was, it would have had to have the wingspan of a Cessna. We both noted how brightly illuminated it was by the moon. We started talking about how a bat's wings are translucent and when backlit from the bright moon, they would appear as we saw them. We could not see the body of the creature.

Around five minutes later, we began seeing lights appear in the South Eastern skies. They appeared only in a small window of the sky. Moving SW to NE, looking star like, they would appear out of nowhere, start off dim, then grow in brightness to be brighter that any star in the sky, then dim back down and then disappear altogether. This happened sporadically for the next 5-10 minutes. There was a total of 15-20 that appeared in the same area of the sky. each one at a time, some 15 seconds apart, some minutes apart, some higher in the sky, some lower in the sky, but all still in the same small window of the sky.

As a reference, we had seen a satellite traverse the entire sky west to east, so we knew there was nothing to explain what we saw. It was a very bright night out with no clouds in the sky."

The witness said that there was no sound involved, and that the creature looked like it was black but with a somewhat transparent area to the top part of the wings, which were bat-like.

Mr. Graham is an experienced FAA licensed drone pilot and stated that he would know the size and altitude of objects flying overhead. He also stated that this creature did not behave as a drone would.

The Starlink Satellites were supposed to be visible over Kansas City that evening at around the same time, however many field investigators were watching at the same time during a scheduled sky watch vi a ZOOM link, and none of us could see the Starlink satellites due to light pollution from the city. Therefore, the lights that the witness described were unlikely to have been Starlink.

Note: Starlink has released hundreds of satellites to date and plans on putting many more into orbit. They appear as bright lights in straight lines as they move above the Earth.

August 4, 2020 8:40 pm Kansas City, Missouri - Huge Flying Creature with Bat-Like Wings

Richard Frieburg, age 63, lives on 58th Street between Ward Parkway and Wornall in Kansas City. This is an area where there are a lot of houses. He was watering tomato plants in his garden when he heard an odd scream-like noise with a duration of 5-8 seconds. It sounded human-like, so he assumed it was kids playing across the street and did not think much of it. Then he heard another strange, horrible scream, then two more. At that point he imagined that someone was mistreating a dog and tried to pinpoint where the sound was coming from. Richard then realized that the sound was coming from a large pin oak tree just outside the fence in his yard and looked up. Even though it was getting dark he could make out the shape of something sitting on a branch.

At that moment, the creature unfurled its wings in an awkward way. It appeared to have elbows that were jointed and when the wings, which had no feathers, were extended they appeared like bat wings. The creature had a thin tapered body which measured approximately three to four feet tall and had a wing- span of approximately five to six feet. The witness did not see the eyes of the creature, which flew off to the east towards Wornall road. Instead of running after it to get a better look, Richard just stood there as if he were paralyzed for a few seconds while the creature few out of sight. He could not explain why he reacted in that manner, and still does not understand it. Richard said that normally he would have followed the creature in order to find out what it was. Richard said that the creature neither looked nor sounded like an owl or other large bird.

Mr. Frieburg related to me that he has had UFO experiences beginning at a young age. This may be significant and could explain the temporary paralysis.

August 5, 2020 2:00 pm. Oak Grove, Missouri- Mothman: I received a report from a man who drives for Uber. He was driving a client west on Interstate 70 at the Oak Grove, Missouri exit at approximately 2:00 pm when he spotted an exceptionally large, winged creature with a human body flying across the road. The shadow on the highway was at least 20' across, according to the witness. The humanoid was flying from north to south, then went out of sight as the witness continued to drive West. He was driving at approximately 70

MPH. The client was sitting in the back seat and did not see the creature.

The skies were partly cloudy, there was no wind, and the temperature was 78 degrees. Oak Grove is a small town 20 miles east of Kansas City, Missouri. There are a lot of trees and fields in this area with many places for an animal to hide. This sighting is unusual in that it was a daytime event.

August 7, 2020 3:55 pm Kansas City, Missouri- MothMan
Reported to me by phone: A man driving south on Interstate I-35 one mile south of the Broadway Exit saw a huge black winged creature with human-like legs flying in the air approximately one mile from his position. The man had to take the exit and lost site of the creature. He could see no features from that distance except the wings and legs but estimated the size to be 25' in width with the wings stretched out. The creature appeared to have a difficult time flapping its wings and remaining airborne, as has been the case in other sightings.

Unknown date, Altamont, Missouri- Pterosaur or MothMan
A witness and his family had been hearing strange clicking and shrieking noises in their cornfield for weeks when their son finally saw what was making the strange sounds. The witness saw an eight-foot tall, winged creature with a long muzzle that resembled the face of an alligator. The creature had no feathers and had grey skin and a wingspan which the witness

estimated at over 80 feet in width. The wings looked like bat wings.

Source: https://cryptozoologynews.com/pterodactyl-in-the-corn-fields-of-missouri-town/

Altamont is located northeast of Cameron, Missouri.

2019-2020 Independence, Missouri – Strange Unexplained Shrieking Sounds

For about a year, residents in Independence, Missouri who live near the railroad cut between Sterling Avenue and the Depot have been hearing strange loud clicking sounds, horrible screams, and shrieking sounds that cannot be located or identified. In some cases, witness report hearing something very large moving in the trees and moving tree limbs, but they can see nothing. At times people have reported thinking that a woman was being murdered, but upon investigating can find no source. One woman heard what she thought was a woman screaming outside her home at 11:15 pm on August 29, and immediately ran outside to investigate, but could see no one. She called out but no one replied. The woman then heard a strange shriek further down the railroad tracks behind her home and called police, who found nothing. She thought that the second shrieking came from an animal rather than a human. Perhaps people need to be looking up in the trees rather than on the ground.

Some investigators have made a connection between rivers and railroad tracks – with the common trait being that of movement. The movement may create portals which allow inter-dimensional creatures to pass through.

2020: Cheyenne Bottoms, Kansas- Pterodactyl Sightings

This creature was native to Kansas and Missouri millions of years ago, back when the Western Interior Seaway covered most of this area with a wide expanse of water. The largest pterosaur fossil found had a wingspan of over 25 feet.

But they may not be extinct. Multiple witnesses have spotted a pterodactyl west of Cheyenne Bottoms in recent years. This site is a wetland occupying 41,000 acres in central Kansas in Barton County and is the largest wetland in the interior United States. At least 340 species of birds have been observed at Cheyenne Bottoms. Perhaps there is one more species the Kansas Department of Wildlife and Parks has not included in that list.

According to investigative journalist Jonathan Witcomb, he has taken four sighting reports of such a creature at different locations in Kansas and made a YouTube video about the reports which can be viewed at https://youtu.be/AgEyzIYb9go.

August 7, 2020 Hannibal, Missouri – Winged Humanoid

A woman by the name of R.A. was driving on interstate 70 under the overpass at the Hannibal exit when she saw a tall humanoid creature with huge wings standing next to the overpass just after dusk. R.A. screamed and slowed the car, waking her husband who had been sleeping. Her husband confirms the incident, although he did not get a glimpse of the creature.

Late August 2020 Cape Girardeau, Missouri –Large Winged Creature

A woman and her sister were driving near Cape Girardeau and looked up to see a large blackish/brownish winged figure flying above the road. She saw a whipping, fluttering motion and something that took up the entire road.

Her sister got a better look at it. She said that the creature was approximately eight feet off the ground and was five to six feet long and had huge membrane-like wings which were ribbed like an umbrella.

Source: https://www.singularfortean.com/news/2020/11/24/sisters-report-sighting-of-what-looked-like-a-body-shrouded-in-wings-near-site-of-previous-report-in-cape-girardeau-missouri

October 10, 2020 Marseilles, Illinois – Flying Humanoid

A group of four men were fishing at the cooling pond near the plant south of Seneca, Illinois at approximately 6:00 pm when they noticed an exceptionally large bird fly over them. The said it was 20-30 feet in the air and was moving silently and barely moved its wings. As it got closer, the witnesses saw that it looked like a tall man with exceptionally large wings. The creature flew out of sight and the men headed home. *From Lon Strickler, ufoclearinghouse.wordpress.com.*

Unknown date – South-Central Missouri: Flying Creature

A strange large flying creature was spotted in the so-called Marley Woods area of South-Central Missouri by a local resident. The witness sad that the wings seemed to be metallic. The witness did get a picture of it and gave it to researcher Tom Ferrario.

The area was named Marley Woods by UFO researcher Ted Phillips, who lived nearby and conducted multiple investigations in the vicinity over many years, but it is not an official name. The residents wish to remain private and do not want other researchers, except those they are currently working with to be on their properties. The area is known for paranormal activity of all types.

November 26, 2020 Oregon, Wisconsin – Person with Huge Wings Spotted by Two Young Women

Claire and Ashley, both age 18, told Investigator Tobias Wayland with the Singular Fortean Society that they saw a large unknown creature as they drove outside of town. They saw something move across the road some distance in front of them at approximately 10:00 pm but could not identify it. Then, at approximately 10:30 pm both women saw a flying humanoid that appeared to be nine feet in length with huge wings swoop down and over a streetlight, which illuminated the underside of the creature. Their first thought was that it was a pterosaur. The women drove on, then decided to go back at 11:00 pm to see if the creature was still around, and to their surprise, one of the women saw something standing in a field. Although it was difficult to see in the dark, one woman described it has having no neck and red eyes, and it was stocky. Each time they drove through the area the car radio had interference and even changed channels on its own.

November 29, 2020 Oregon, Wisconsin – Winged Creature Returns

Claire (continued from the above story) again saw the winged creature above as she drove on the same road. First, she saw a shadow on the road and knew it was the creature, then she saw it run into the trees. She could tell that it had wings and was large.

December 2020 Weldon Spring, Missouri – Giant Bird Shadows Car

Donna Wilder was driving on Wolfrum road in the small town of Weldon Spring on a clear evening when she saw a gigantic flying creature fly low over the road. It appeared to be following the creek and woodline. She said, "It was extremely large and the thing completely shadowed my car." Wilder did not stop as it scared her to death. Weldon Spring is in St. Charles County, Missouri near the Mississippi River.

December 2020 Independence, Missouri – Huge Winged Humanoid Near the Missouri River

Tony Degn was driving for Uber one evening and headed to the Kansas City Airport to pick up a client. He stopped to get his car washed and get gas. Next, he picked up his client, who happened to be a player for the Kansas City Chiefs. As Tony drove south on I-35 highway the client commented on the fact that there were very few cars on the highway. The two discussed this, as this highway is normally busy at all times of the day or night and they found this to be very strange. There was no accident so they could not explain this. Tony also had a strange feeling that the Mothman being was nearby again.

119

Since Tony has seen this creature several times, he now has a certain indescribable heavy sensation occur when it is near him. However, neither he nor the client saw anything strange, or at least they do not remember it. After dropping the client off at his home Tony headed back and realized that he had an hour and forty minutes of missing time that he could not account for. Tony then headed home and got out of his car and that is when he noticed a large amount of a strange substance on the roof of his vehicle. It was late, so the next morning he I examined the reddish substance on the vehicle, which was in multiple areas in round splotches. It was also crystallized. I took a sample of the substance and sent it to Lynne Mann at the Missouri MUFON Lab in Farmington, Missouri. Lynn tested the substance, and it is urine. She also tested the pH level, and the result matches a bat. I showed the photos to our FI team and everyone agreed that whatever it was that urinated on Tony's vehicle was big due to the large amount of urine. We are currently trying to obtain DNA from different types of flying creatures that could have possibly been connected to this case and when I get the results they will be posted on my website.

TOP OF VEHICLE PHOTO: MARGIE KAY

CLOSE UP OF ONE SPLOTCH

Tony has since driven over the Liberty Bend Bridge on 291 Highway on numerous occasions and has felt the presence of the mothman creature several times in this area. For this reason, Tony believes that the creature may be hiding under the bridge.

This area is wooded and would make a good habitat for any animal. The fact that mothman has been connected to a number of bridges cannot be ignored. They may actually sleep underneath bridges during the day and come out at night, which is when most sightings occur.

LIBERTY BEND BRIDGE
BY AMERICASROOF AT EN.WIKIPEDIA - PHOTO BY POSTER, CC BY-SA 2.5,
HTTPS://COMMONS.WIKIMEDIA.ORG/W/INDEX.PHP?CURID=16224270

The Disappearing Crane Incident in Broken Arrow, Oklahoma

Thank you to Mindy Tautfest, MUFON State Director for Oklahoma for this account:

"During the spring of 2020, I attended a sky watch with a gentleman from one of my former MUFON cases. He and I had stayed in touch after the completion of his case and he regularly shared new photos and videos he had taken of different craft and lights in the sky. He had been known to have the ability to call down UFOs from the sky and had indeed produced several interesting photos of unidentified craft to back up his claims. The night of May 29th afforded clear skies with great visibility when we all gathered up on that hill located in what was once part of the old Osage Nation Reservation land. The hill was rocky with wild sagebrush growing in patches around large collections of boulders. Nearby city lights combined with the brightness of the half moon and illuminated the area well enough for us to see our footing as we ascended the small mound. Once atop, the vantage point gave us a perfect view of the city and surrounding areas below including the nearby airports with a clear sight of the flight patterns of planes taking off and landing.

There were a handful of us in attendance for the duration of the sky watch, with a few others coming and going throughout the night. Those who were in continuous attendance were me, the witness, his colleague, and two MUFON members; one of whom is also an experiencer, the other who opted to stay with the car at the bottom of the hill due to the strenuous climb required. After several hours of viewing the sky, we had seen 3 questionable lights which were not easily identifiable as known commercial craft. Not a bad evening, but honestly, nothing quite mind blowing either. All the lights we had seen had been far in the distance and could have been unmarked military craft, drones, etc. since none had displayed any irregular maneuvers. As it was nearing midnight, we decided to call it a night and began packing up our gear in preparation for the downhill hike which lay ahead. As each of us was getting equipment torn down and packed away, our conversation briefly turned to the correlation between UFO sightings and birds. It was a phenomenon which both experiencers in attendance had encountered numerous times before. I listened as they each swapped stories about redbirds, owls, cranes, and black birds, all seen in the presence of other unexplained phenomena. Some view them as messengers. Others think of them only as screen memories, but one thing was certain, birds were a recurring theme known to those who have had UFO encounters.

We stood at the precipice of our descent and wrapped up our conversation as we prepared for our downward trek. Just as we turned to head down the hill, a sandhill crane appeared at the level of a nearby power pole. The pole had lines coming off of it at approximately 15 feet above ground level and we stood about 20 feet away from where the pole itself was planted. When we noticed the bird, it was nearly directly overhead. It flew between the powerlines before swooping down to 10 feet overhead, and then heading off in an easterly direction. As it passed over, we could easily see the details of the bird. White and tan feathers were interspersed and covered the body. Individual feathers were visible along the tips of its wings which boasted a wingspan of close to 7 feet. It carried it is neck in a folded fashion and its two thin legs hung down below its body as it flapped its wings and soared up to 80 feet. We all stood and watched as it reached the height of its aerial climb and in a quite spectacular display, simply blinked out of existence.

CRANES IN FLIGHT IMAGE: ADOBE STOCK

There was a palpable and collective gasp of awe as each in our group struggled to make any sense of the unbelievable spectacle, we had all just witnessed. We waited in silence for several more minutes in anticipation of a reemergence, but the creature never returned."

Strange Giant Bird, UFO, and Fairy Creatures in Oklahoma 2019-2021

Mindy Tautfest, State Director for Oklahoma MUFON put me in touch with a couple in Oklahoma who saw a Pterodactyl type creature and have been seeing strange fairy-like beings on their property and nearby. The following is information I obtained from the witnesses directly:

Pterodactyl

In 2019 Deborah and Lloyd M. headed to church one early morning from Midwest City to Norman. They drove on Sooner Road where there are farms, cattle ranches, and woods. Deborah noticed something just above the tree line to her west and pointed it out to her husband. The two watched a 10-foot-long brown/gray bird with bat like wings flying lazily up and down close to the tree line. The creature headed downward and disappeared. The couple said that there were no feathers visible on the creature and they heard no sound. They estimated the wingspan to be at least 20 feet. It then passed ahead of them and headed east towards Stanley Draper Lake

and went out of sight. The couple said that it resembled a pterodactyl more than any other type of bird.

Slenderman?

After a severe ice storm which severely damaged or killed at least 80% of the trees in the area the couple had two very frightening experiences. The first event occurred while they were outside taking pictures of the stars. Deborah and Lloyd both suddenly had a feeling of dread and became scared. Deborah noticed something move between two stacks of felled wood on their neighbor's property. The creature moved oddly in a slithering fashion. While crawling on the ground looked like a large reptile with a tail, but when it rose up it grabbed onto a tree with two arms and extended itself upward to approximately 15 feet in height, while becoming very slender at the same time. The couple described it as a fuzzy tall stick man or something like the so-called "Slenderman." Deborah and Lloyd both became very frightened at that point and held their breath as the creature took one long 30' step into a creek and disappeared. At that point, the couple ran into their house.

The second frightening experience occurred not long after while they were once again in their back yard. They again had a sudden feeling that something was not right. They heard the fence move loudly and they looked in that direction. Something with long limbs and body, which they estimated was at least eight feet tall jumped one fence in a single bound, then moved through the neighbor's yard in just two steps. The creature went behind the neighbor's shed, then disappeared. It scared them so much that the couple did not go outside for several days. However, they have not experienced a feeling of dread again and they feel comfortable on their property. Perhaps the Slenderman legend comes from a sighting of a creature such as this.

AREA WHERE THE PTERODACTYL-LIKE CREATURE WAS SPOTTED. NOTE THE FARMLAND, WOODS, AND LAKE WITH AN ABUNDANT FOOD AND WATER SUPPLY.

UFO and Fairies

In February of 2020 the couple was again sky watching and noticed a strange object in the sky. They wanted to get a closer look at it, so they viewed the craft through binoculars. They could see something that looked like molten mercury emitting black objects which went up and then back down, and white lights which fell from the object to the ground. The UFO

128

report was filed with MUFON, but then the couple began to experience something even more bizarre when they started to see small 6" tall humanoid creatures flying around their property and nearby.

One day soon after the UFO sighting Lloyd was outside in his yard and noticed a small 6" tall creature flapping its wings. It went to the oak tree on the property. Lloyd said that it looked like a combination of a human with a squarish head and a bug. It had long skinny arms and legs and big feet with three or four claws on the front of each foot and one on the back. It had oval shaped eyes that came together in the center. The creature was kicking its feet as it flew. It had yellowish color legs and arms and a darker body. The wings were small and white like goose feathers. Unbelievably, the creature smiled at Lloyd. As Lloyd moved in towards it the creature flew up into the tree.

Not long after Lloyd's encounter Dorothy would have her own. She was driving at approximately 2:00 pm in the afternoon not far from her house when she noticed something fly out of a tree towards the road. At first Dorothy thought the creature was a sparrow, but as it approached, she realized that it was something else. Dorothy stopped her car and got out to get a better look, and the flying creature flew just 10 feet from her and 10 feet above her. It then hoovered, staring at her. The creature looked female and had legs with feet like bird claws, arms, and a human head with eyes that were more like bug eyes. The creature blinked both up and down and side to side like a reptile. She had a triangular shaped brownish/taupe color body at the waist which Dorothy described as looking like a skirt; wings; and it was carrying a bundle, which reminded Dorothy of a baby wrapped up in a blanket. The two just stared at each other, then when Dorothy moved closer to get a better look the tiny creature flew off very quickly.

Since these events Lloyd and Dorothy have encountered the tiny, winged creatures on numerous occasions, and sometimes they even swarm around Lloyd's head. One day as Lloyd was working outside, he heard a distinct telepathic message to "turn around and look." He did so and was shocked to see several fairies sitting or lying on a Mimosa tree branch. The couple said there are different sizes of these creatures and they fly three times as fast as a sparrow. Once they saw one pass a mockingbird, and apparently that would be quite a feat. The couple has been able to capture a few of these creatures on camera and they both made drawings of two of the creatures they saw. They call these creatures Mimics, although they do not know why.

Both Dorothy and Lloyd are intuitive. After Dorothy had a near-death experience in 1993 after receiving a severe electric shock, she realized she had some psychic abilities. One has to wonder if the creatures are not attracted to the two of them due to their level of awareness.

As an investigator, I believe we need to look at the first event, which is the couple's sighting of a UFO which spewed out a substance or lights. This event preceded the Pterodactyl sighting, the Slenderman sighting, and the subsequent tiny fairy sightings which continue to this day. Is there a relationship? If so, are these strange creatures actually from another world or dimension?

FARIES OR "MIMICS" PHOTO BY DOROTHY M. ENHANCED BY MARGIE KAY

THESE CREATURES ARE VERY CLOSE TO THE CAMERA AND ARE ABOUT 6" TALL.

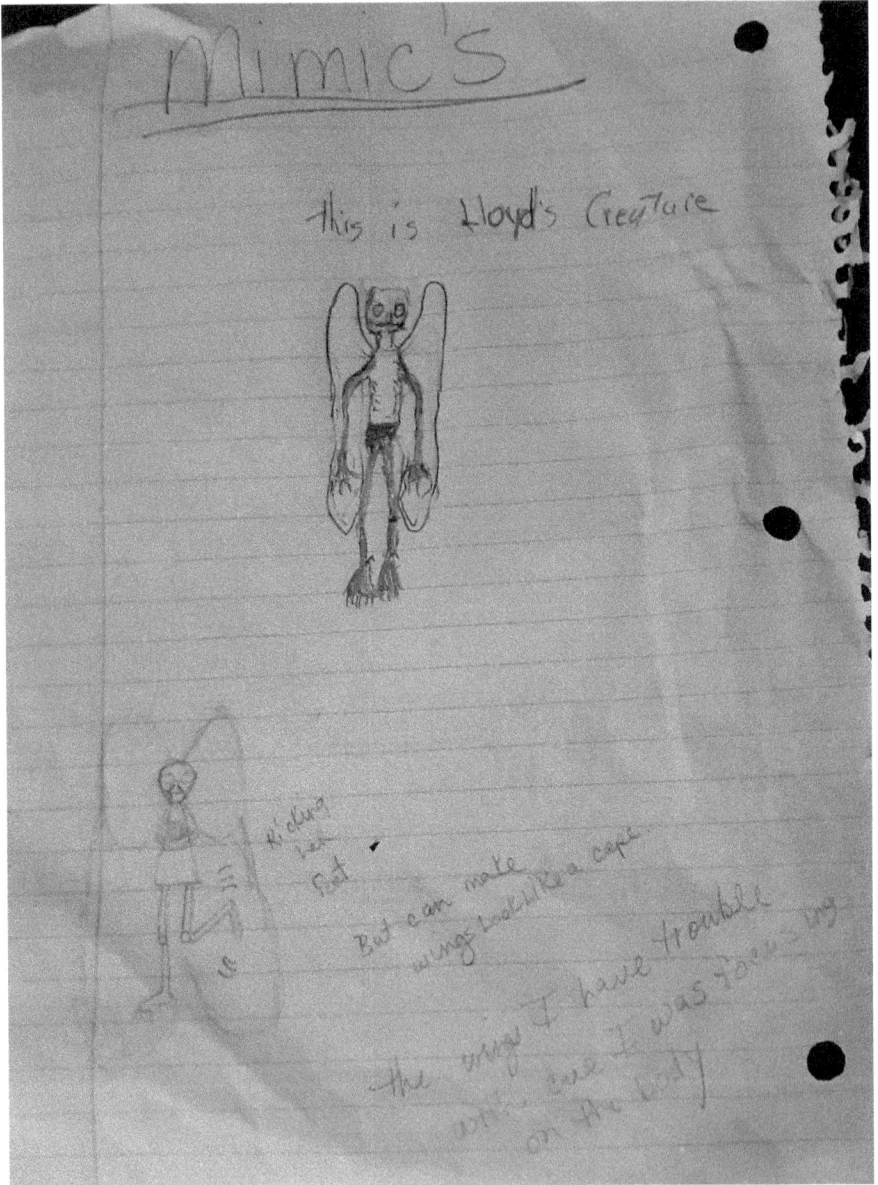

Mimic's

this is Lloyd's Creature

Kicking leg. Fat.

But can make wings look like a cape.

the wings I have trouble with... (one I was focusing on the body)

DRAWINGS BY LLOYD AND DOROTHY M.

Central Alabama Fairies and Other Elementals

Ron Johnson began seeing fairies and other elementals at his home in Alpine, Alabama in 2003 and still sees them today. Ron has also seen something less than a foot tall in the shade that looks like a gnome, as well as clouds of glittery sparkle at the edge of the woods, and trees full of lights that are much larger than lightning bugs that remind him of Christmas lights in all colors. He even saw one entity that is at least 30 feet tall and is pencil thin. The entity is made of light (a light being) and Ron has seen him four times in a neighboring pasture. All of the sightings were daytime sightings except for the colored lights. Ron is a Native American Shaman and a very spiritual person who meditates on a regular basis. It is my opinion that as a person raises their vibratory rate, they find it easier to see into the fifth dimension. Ron agrees.

February 12, 2021 Independence, Missouri – Interdimensional Fairy

This is my own sighting. I was sitting in the hot tub on our deck at approximately 10:30 pm and saw a flash of light that looked like a circular window and a 6" tall fairy with a humanoid body and wings come out of the portal, fly across the deck for approximately 10', then another flash and portal opened, and the creature flew into it and disappeared. The fairy was glowing gold with a white aura around it and it kept looking straight ahead as it went into the second light. I was astounded by what I saw and realized that this was a creature traveling between dimensions. I can think of no other explanation. I asked Kristina McPheeters to draw what I saw, and this is a very close representation:

133

GOLD FAIRY AS SEEN BY MARGIE KAY DRAWING BY KRISTINA MCPHEETERS

I find it very coincidental that while I was working on a book about winged creatures which includes fairy-like beings, this event occurred right in front of my eyes. After all, it has been many years since I have seen anything like this. What was the purpose? Was some intelligence showing me that these beings move between dimensions in order to help me solve the mystery?

March 27, 2021 Weldon Spring, Missouri – Winged Alien?
Donna Wilder took video of a rainbow right after a hailstorm from her home in Weldon Spring at 5:11 p.m. on March 27, 2021. This location is near rivers and airports. She used a new Samsung S20 Note Ultra phone to take the 31 second video. While reviewing the video frame-by frame she found what looks like an exceptionally large bird that looked out of place. She also found white lights under the rainbow that she cannot explain, and multiple high-speed UFOs. Donna was on the lookout for these type of craft after reading the book "The Fast Movers: Evidence for High-speed UFOs/UAPs". We also noticed that in the first photo the sky is noticeably lighter under the rainbow than the rest of the sky with a big difference in color, and we have no explanation for that.

After reviewing this on her computer screen, Donna said she thinks the creature has legs behind it. These photos were analyzed by Wayne Lawrence, a video analysis expert. Wayne found multiple groups of high-speed UFOS which he calculated to be moving a 6,600 Mph! The winged creature moves too fast and in odd ways to be a normal bird, and it is too large. I researched several bird websites and compared silhouettes and descriptions but could find nothing that matched the shape of this creature.

What Donna captured on film may finally be definitive proof that some winged creatures are actually aliens or alien craft!

FRAME FROM VIDEO OF LARGE FLYING BIRD/CREATURE   PHOTO: DONNA WILDER

CLOSE UP OF FLYING CREATURE   PHOTO: DONNA WILDER

STILL IMAGE FROM DONNA WILDER'S VIDEO SHOWING UFOS

There are six UFOs in this single frame from the video showing the direction they are traveling. Hundreds of UFOs were captured in Donna's video – along with the winged creature!

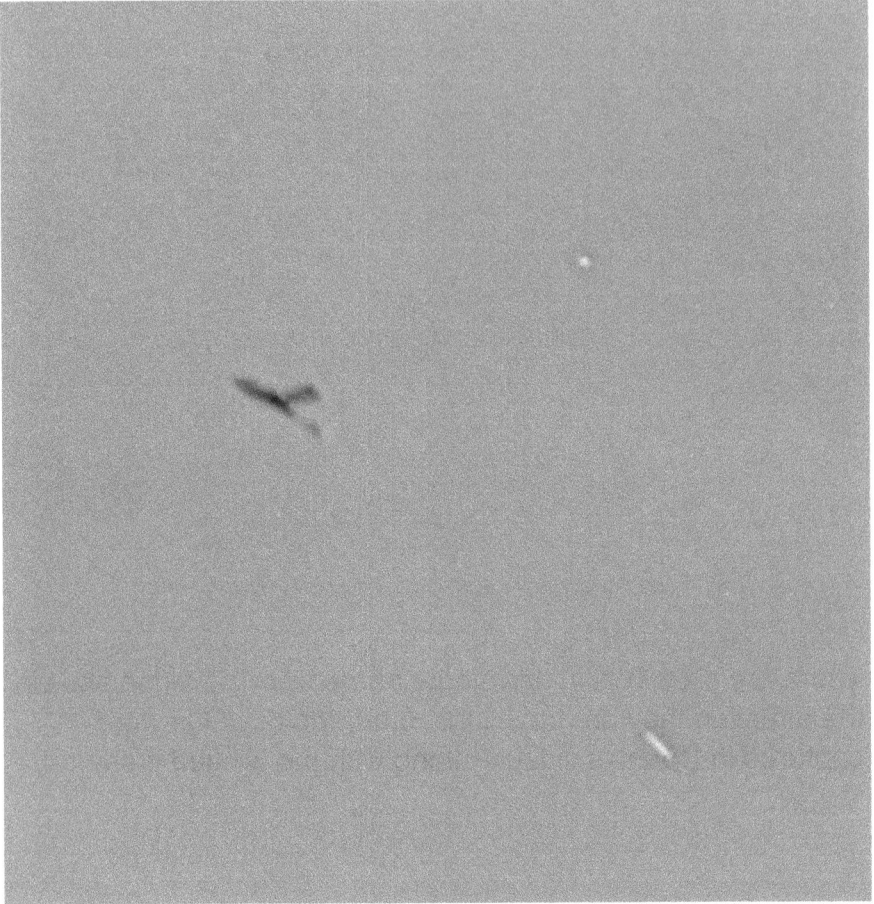

THIS SINGLE FRAME SHOWS WHAT LOOKS LIKE A WINGED CREATURE ALONG WITH TWO UNIDENTIFIED OBJECTS WHICH DONNA SAW AS LIGHTS PHOTO: DONNA WILDER

Winged Aliens

The following frames from the video were extracted by Wayne Lawrence:

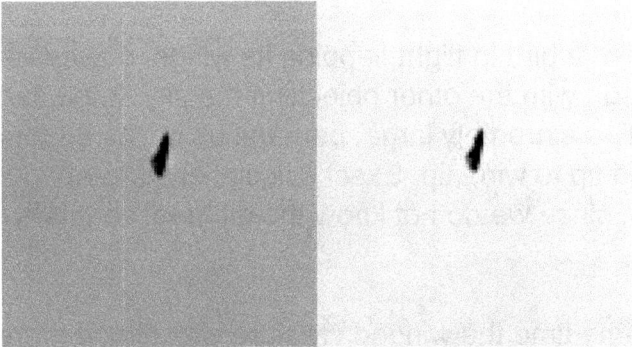

Frame 18:05   Enhanced

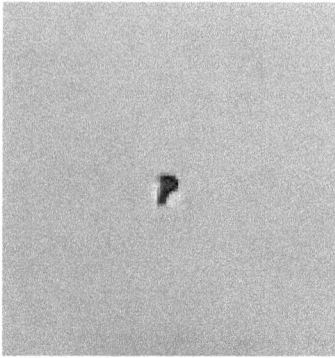 

Frame 18:07 Enhanced

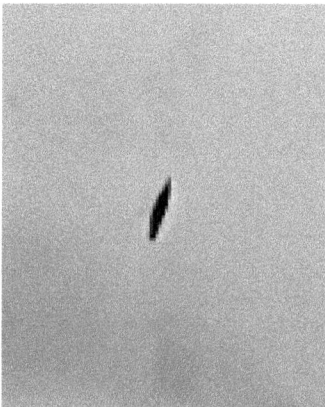 

Frame 18:10 Enhanced

It looks like a bird in flight flapping its wings, however, in comparison with the other objects in the sky at the same time this "bird" is extremely large, perhaps as big as 88 feet in width from wing tip to wing tip. Exact calculations are not possible, however, since we do not know the altitude, so this is a best guess.

At the same time the winged creature was filmed there were multiple UFOs captured on the video at the same location. Below is one of the better photos of one of these objects. It clearly is not a conventional aircraft. The object was 22.3' in length.

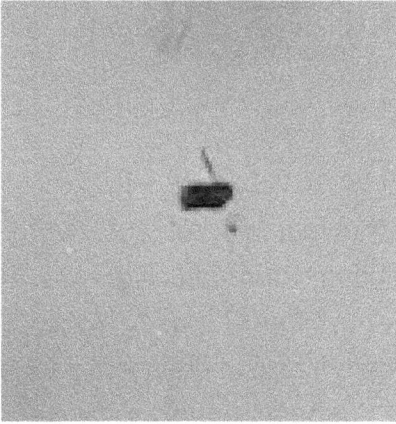 

Frame 27:114   Enhanced

Could the UFOs have anything to do with the winged creature or craft? Could the winged creature actually be something other than a bird?

April 2, 2021 Southeast Kansas - Winged Creatures Inside Balls of Light

A family in southeast Kansas has been experiencing many strange anomalies for several years. This site is currently under investigation by Kansas MUFON. Among other things, the witnesses have seen small balls of lights with something inside of them that look like winged creatures. While this is very unusual, it is not unheard of. The family has seen UFOs and other phenomena at this site, and they believe there is a vortex at this location. The witnesses say it is something like the Skinwalker Ranch and that there is a portal located on their property. The correlation between UFOs and the winged creatures inside orbs cannot be discounted.

OTHER SIGHTINGS AND PHOTOS

Mars Creature

A recent photograph from Mars shows a large object which is airborne and appears to be a flying creature of some type. This picture was taken by NASA from the Mars Curiosity Rover and the area is the Gale Crater. Marcelo Irazusta first noticed the flying bird in the picture. One cannot help but notice the shape, which resembles a Mothman, bird, or winged aircraft. If it is indeed a Mothman, this would support the theory that this creature is alien.

OBJECT OVER MARS PHOTO: NASA & EXPRESS.CO.UK

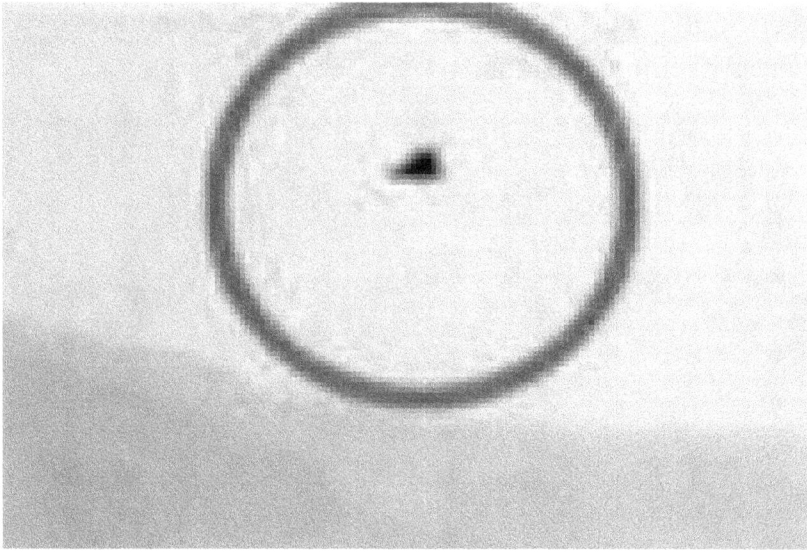

CLOSE UP PHOTO: NASA & EXPRESS.CO.UK

Flying Humanoids in Mexico

Since the year 2000 there have been a number of reports out of Mexico which include flying humanoids both with and without wings, and even "flying witches" or "bruhas." These reports have been covered by a number of journalists and have appeared in several TV documentaries and online blogs.

On October 1, 2000, the La Prensa newspaper from Mexico City published a report from a commercial airline pilot and his co-pilot who witnessed a flying man during their flight. According to the article: "A commercial airline pilot from Aero California who wanted to omit his name to avoid problems in his work reported the sighting of a "little flying man" who was flying at the same altitude of the plane before landing. According to the pilot this "flying man" had a kind of backpack

143

in his back and was flying freely. The pilot added he saw perfectly well arms and legs.

En septiembre

Más reportes por aparición de OVNI's en el país; hasta un hombrecito volador

PABLO CHAVEZ

Durante los primeros días de septiembre se reportaron varios avistamientos de Objetos Voladores No Identificados (OVNI's) sobre la ciudad de México. Los casos quedaron videograbados por aficionados, quienes también reportaron la presencia de un "hombrecito volador" por las inmediaciones de ciudad Nezahualcóyotl.

Por su parte un piloto de la aerolínea Aerocalifornia, quien se reservó su nombre por temor a represalias en su trabajo, reportó el avistamiento de un "hombrecito volador". Ese ser, dijo, traía una mochila en su espalda. El piloto externó que vio claramente sus extremidades superiores e inferiores.

144

Old Photo of a Huge Flying Creature

PTERODACTYL LIKE CREATURE HUNG ON DISPLAY. SOURCE: UNKNOWN

The photo above emerged in 2011 but the source and location are unknown. The photo looks old, perhaps from the 1940's or 1950's. The beak looks like that of a crane, the body has feathers, and the wings seem featherless. The scene seems genuine. Could this be a Thunderbird or Pterodactyl? Until the source comes forward, we may not get answers, or find out if this is real or faked.

Possible Mothman Sighting at a Bridge in Ohio

The following is a widely circulated photo. It was taken in 2003 by someone at the Ironton-Russel Bridge which crosses the Ohio River between Russell, Kentucky and Ironton, Ohio. In the second photo, the creature flies off the bridge. There is a lot of controversy about this incident. Some people claim that the object is a piece of metal that fell off the bridge. I could not locate the person who took the photos. These are circulated widely on the internet.

FIGURE 2 POSSIBLE WINGED CREATURE ON THE BRIDGE

FIGURE 3 CREATURE LEAVING THE BRIDGE

CONCLUSION

The UFO Connection

There is sometimes a connection between UFO sightings and winged humanoid or winged creature sightings. The June 20, 2019 sightings of winged creatures occurred on the same day as a mass UFO sighting over Kansas City where thousands of people witnessed three stationary white objects for over six hours. Was this simply a coincidence or is there some connection?

During the Mothman scare in Point Pleasant, New Jersey people also saw many Unidentified Flying Objects which were more plentiful than the Mothman sightings. Some people were visited by strange Men in Black, and many researchers such as John Keel and others speculate that there is a relationship between the UFO sightings and the Mothman sightings. One must wonder if these humanoid flying creatures are in fact from Earth or elsewhere.

To date, I have taken over 70 reports of giant winged creatures in the state of Missouri and quite a few from other states, with most sightings occurring near the Missouri River. Several sightings have occurred just before or after UFOs were spotted in the same locations. In total, I have collected over 100 reports taken by me and others.

In one location at Sugar Creek, Missouri, the witness has seen multiple unidentified flying objects, orbs, an alien in her yard and in her house, and a giant bird with a huge wingspan that was larger than a small plane. She also watched a hawk morph into a UFO. This location seems to be a paranormal hot spot, and it is very close to the Missouri River.

In other cases, witnesses reported seeing balls of light just prior to or after seeing a gigantic, winged creature with bat-like wings. Balls of light have often been associated with UFOs.

Could there be a connection between the winged creatures and alien craft? It begs the question: Are these winged creatures aliens rather than Earthly beings?

The Habitat

In doing research I have found that most of the sightings of humanoid flying creatures and gigantic birds were in Kansas City, Missouri/Kansas and Chicago, Illinois with a few scattered in different areas across the United States. What is it about these areas that the creature is attracted to? Perhaps it is water, which both cities have plenty of.

It may be possible that there is a larger colony of these creatures in the Chicago area, and they followed the Illinois River to the Mississippi River and the St. Louis area, then followed the Missouri River to Kansas City. This is just speculation of course, but large creatures such as these would certainly need a water source. Food sources nearby and in the rivers would be abundant as well. It is interesting to note that 95% of the sightings occurred within one or two miles of a major water source.

RIVERS AND CREEKS IN THE KANSAS CITY AREA

PUBLIC DOMAIN, HTTPS://EN.WIKIPEDIA.ORG/W/INDEX.PHP?CURID=10190217

Swope Park is the 51st-largest municipal park in the United States, and the largest park in Kansas City. It also the central hot spot in the city for winged creature sightings. The park is named in honor of Colonel Thomas H. Swope, a philanthropist who donated the land to the city in 1896 (coincidentally, a cousin of mine). The park contains, lakes, streams, woods, fields, The Kansas City Zoo, Starlight Theater, and other sites. Food and water sources are abundant.

A number of giant bird and Mothman sightings as well as Sasquatch and UFOs have occurred in or near the park. Perhaps there are cryptid residents of this hilly wooded area that are as yet unknown to the general public.

Swope Park, Kansas City, Missouri with Starlight Theater in the foreground
Photo: CC BY-SA 3.0,
https://en.wikipedia.org/w/index.php?curid=13326228

This Google Earth view shows the locations of winged creatures, UFOs, and two Sasquatch sightings, all located in or near Swope Park

KNOWN SPECIES

The only known fossil of *P. sandersi* was first uncovered in 1983 at Charleston International Airport, South Carolina, discovered by James Malcom, while working construction building a new terminal there. At the time the bird lived, 25 million years ago, this area was an ocean The bird is named after Albert Sanders, the former curator of natural history at Charleston Museum, who led the excavation It currently sits at the Charleston Museum, where it was identified as a new species by Dan Ksepka in 2014 "Though no feathers survived, Ksepka extrapolated the mass, wingspan and wing shape from the fossilized bones and fed them into a computer to estimate how the bird might fly. A conservative estimate put the wingspan of *P. sandersi* at around 6.4 meters (21 feet)."

Pelagornis sandersi is an extinct species of flying bird, whose fossil remains date from 25 million years ago, during the Chattian age of the Oligocene. The sole specimen of *P. sandersi* has a wingspan estimated between 6.1 and 7.4 m (20 and 24 ft), giving it the largest wingspan of any flying bird yet discovered, twice that of the wandering albatross, which has the largest wingspan of any extant flying bird (up to 3.7 m (12 ft))

In this regard, it supplants the previous record holder, the also extinct *Argentavis magnificens*. The skeletal wingspan (excluding feathers) of *P. sandersi* is estimated at 5.2 m (17 ft) while that of *A. magnificens* is estimated at 4 m (13 ft)

1m

Pelagornis sandersi comparison with the Andean condor (*Vultur gryphus*) and the wandering albatross (*Diomeda exulans*)
CREDIT: ANAXIBIA, WIKIMEDIA COMMONS

Argentavis magnificens

Photo: Dr. Kenneth E. Campbell with the 25 ft. wingspan of Argentavis magnificens. Display from the Natural History Museum, Los Angeles

Argentavis was big. Real big. Big enough for National Geographic to compare the huge raptor to a Cessna 152 two-seater airplane. It averaged a 70 kilograms (over 150 pounds), and its wingspan measured nearly 23 feet from tip to tip.

Bats

The largest bat in the world is the Giant Golden-crowned flying fox, which can weigh 3.5 lbs. and have a wingspan of 5.7", however these bats do not live in North America. The largest bat in the Midwest region is the giant brown bat, whose wings only reach up to 13.5". Therefore, we can rule out any known bat species.

Other Known Giant Birds

The wandering albatross has the largest known wingspan of any living bird, at times reaching nearly 12 feet. But millions of years ago, there was a bird with wings that dwarfed those of the albatross, researchers now report. The newly named species, Pelagornis chilensis, which lived about 5 million to 10 million years ago, had a wingspan of at least 17 feet, and the largest pterosaur fossil ever found had a wingspan of 25 feet in width.

Several witnesses mentioned in this book described gigantic birds with wingspans ranging from 10 to 36 feet. Some speculate that the flying bird is a Thunderbird or Pterodactyl.

The second largest known living bird in the United States is the California Condor, which has a wingspan up to 9.8 feet. The

Condor's habitat is not in the Midwest region, so it must be ruled out as a possibility.

I submitted an inquiry to the Missouri Department of Conservation and here is their reply:

"The largest bird wingspan in Missouri is a bald eagle at 80" (just shy of 7 feet). A bald eagle would be fairly easily recognizable and would not perch on a car but soar high above or seen perched at a distance. A turkey vulture would, however, perch on a car and has a wingspan of 67" (about 5.5 feet). Not large enough to cover an entire car, though.

No bird in Missouri has a 20-25 feet wingspan. The largest wingspan of any flying bird is the critically endangered California Condor, only found in tiny pockets in the far west."

Jonathan Whitcomb states in his book "Live Pterosaurs in America" that approximately 150,000 Americans have had significant sightings of pterosaur or bird or bat-like large flying creatures. Although no living or dead specimen has been captured, at least in recent years, this is a significant number of sightings. And these are only the numbers of persons who filed a report with someone. Obviously, there is a case for the possibility that the pterodactyl did not die out 65 million years ago, and a few may be living today, and that there are other types of creatures in existence which are yet unidentified.

Turkey Vultures have large wingspans up to 72 inches and could be mistaken for the legendary thunderbird.

A Turkey Vulture (Image: Dolovis/Wiki Commons/Public Domain)

The Condor is the largest known bird in North America and may be the culprit of some giant bird sightings. Its wingspan can measure up to 9.5 feet. Its habitat is the western coastal mountains.

California Condor (Image: Gregory Smith/Flickr)

Ley Lines and Portals

One point I find interesting is that most of the sightings of these creatures in Missouri are focused in the greater Kansas City Area which is known as a major UFO hotspot (see my book The Kansas City UFO Flaps), and as a paranormal hotspot with hundreds of haunted sites as well as other strange phenomenon such as time anomalies and sightings of strange creatures such as giants, bigfoot, ghosts, and fairies.

There are several intersecting main ley lines in the Kansas City area, as well as what we have dubbed "paranormal highways" which are Highway I-35, Highway 49, and 39th Street in Independence, Missouri. These locations have the bulk of unexplained phenomena. According to Wikipedia, "Ley lines are hypothetical alignments of a number of places of geographical interest, such as ancient monuments and megaliths. Their existence was suggested in 1921 by the amateur archaeologist Alfred Watkins, whose book 'The Old Straight Track' brought the alignments to the attention of the wider public." Many of the worlds megalithic structures were constructed on Ley lines.

The sites of many sightings in Kansas City also follow latitude and longitude at 94 degrees W intersecting with 39 degrees N. I have also noted that the numbers 35 and 39 come up often during investigations with relation to not only lines of latitude but street addresses and house numbers as well. It is anyone's guess as to why this occurs, but it is a strange coincidence. Perhaps these sites create portals to other dimensions, which allow non-earthly creatures to travel through, if only for a short time. This may be an explanation as to why no one, at least in recent years, has captured a specimen. Or perhaps they have, and it has been covered up.

Portals are theorized to exist where energy concentrates and flows. This could be a waterfall, a river, a chimney, or even fault lines. High heat can generate a portal, as can an area where magnetic energy is concentrated. A portal can be created by man by building circular or spiral structures or gathering with many people to move around in a circular fashion. Native Americans are known to do spirit dances, which open a portal to other dimensions so their ancestors can more easily move through to this world. In many instances, the ancestors are visible to the living and dance with them. I have watched this myself, so know it does occur.

As for structures, one of the more obvious is the Community of Christ Temple in Independence, Missouri. This structure has created a spiral vortex, and likely a portal where interdimensional creatures can move in and out of easily. The building was constructed in a spiral shape. There have been a large number of UFO, alien, and strange creature sightings within a one-mile radius of this structure.

When checking the area within three blocks of the building using dowsing rods, over 25 people have obtained the same results, which are that the rods indicate that there is a clockwise and counter-clockwise spiral vortex happening at the same time. Counter-clockwise movement is also known as a vile vortex. Energy can move both out and up and in and down at the same time. This is highly unusual, as most vortices are either one or the other, not both. Could this structure have created a portal between dimensions?

COMMUNITY OF CHRIST TEMPLE IN INDEPENDENCE, MISSOURI
PHOTO: WIKIMEDIA COMMONS

Government Interest

Apparently, the government does have some interest in these creatures as evidenced by some witness accounts of surveillance. In one particular case, a witness has seen several giant winged humanoids since 2012, and he is followed by government vehicles, sometimes as many as four or five at a time right after such a sighting.

As for me, the government vehicles appear at my house or follow me when I am working on a big UFO or cryptid case, and my phone and internet is tapped as confirmed by Comcast. Apparently, someone is interested in these cases.

What the government knows about the phenomena, however, will likely be kept under wraps, just as they do regarding UFOs. I would imagine this would be especially true if these winged creatures are aliens rather than a living fossil or other natural phenomena.

The Occult Connection

Sam Maranto, State Director of Illinois MUFON, and his team of investigators have noted some strange correlations in the Mothman sightings they have investigated in the Chicago area. One odd fact is a connection to the word OZ. Some of the sighting reports have been at or near OZ park in Chicago, which is named after the Wizard of OZ series of 13 books written by Frank L Baum, who lived in the area. Oddly, Kansas and Kansas City, Kansas, and Kansas City, Missouri are connected to the OZ books as well. In some of the reports provided, the witnesses used terminology related to the OZ books, which may or may not have been intentional. The

Illinois investigators also noted a strong connection with numbers, in this case, the numbers 13 and 31 keep appearing in the Chicago cases, as the numbers 39 and 35 do in the Kansas City area. There are Masonic connections and symbols involved as well. The Illinois team spoke at a meeting for Missouri MUFON recently and this prompted me to dig further into the numbers and symbols in local cases in Missouri and Kansas.

Common Traits

In reviewing the sightings covered in this book, I found common traits among the witness accounts. In order to understand more fully what this phenomenon is we need to consider these traits. During a sighting, many witnesses report the following:

1. A feeling of dread, anxiety, or that something is just not right.
2. The creatures seem unafraid of the witness and may even approach them.
3. The creatures seem to make no attempt to conceal themselves in many instances.
4. A sensation that time has slowed down or the witness experiences missing time or altered time.
5. The winged creature morphs or changes shape.
6. In most cases only a single entity is seen (in only two cases were two creatures seen at the same time.)
7. The winged creatures' eyes are either black or red. No other color is ever mentioned.
8. In no case has a living winged creature been captured.

9. Many witnesses have also seen Unidentified Flying Objects or lights prior to, during, or after an anomalous winged creature was spotted. In many cases, the witnesses are long-time UFO experiencers.

10. Most witnesses report hearing no sound. In only a few a wing flap is heard, and in still fewer, a very loud thunder-like sound is heard.

Based on the information obtained from witnesses and my own personal observations I believe that there could be several explanations for these sightings of strange flying creatures.

The most obvious is that some of the giant birds are remnants of prehistoric creatures that are known to have existed and thought to have become extinct but in fact are not. But this alone does not explain the fact that no physical bodies have been captured - at least in recent times.

Another explanation could be that the flying creatures are drones, especially in the case of the smaller flying creatures, but drone technology is fairly recent so would likely not account for the earlier sightings.

The most plausible answer, in my opinion, is that due to the behavior of some of the winged creatures there may be some type of supernatural or inter-dimensional beings that transcend space and time. In my own sighting of a gigantic, winged creature and three smaller ones, they simply disappeared without making a sound. In the case in which a witness saw wings enveloping a vehicle, they dissolved into nothing rather than flying away. There are other examples as well. I can think of no other explanation for this type of phenomenon than that these beings are inter-dimensional and can move between dimensions at will, much as the Sasquatch and other unusual creatures are often reported doing, not to mention UFOs.

Theories about dimensions other than our own have been in existence for a long time. Experimental psychologist Gustav Fechner under the pen name "Dr. Mises" wrote a short story titled "Space has Four Dimensions" which was published in 1846. In the book Fechner describes a "shadow-man" who is trapped on a two-dimensional surface and can only communicate with other shadows like him. The story is similar to Plato's "Allegory of the Cave" presented in *The Republic* (c. 380 BC).

In 1898 Simon Newcomb wrote an article titled "The Philosophy of Hyperspace," in which he theorizes that there are higher dimensions.

According to Ethan Stephans of Still Unfold.com, superstring theory may also "…be the Theory of Everything that unites the otherwise incompatible theories of general relativity and quantum mechanics into one insterally consistent collection."

The Theory of Everything was first presented by Albert Einstein. According to physicists, there are at least 10 dimensions as follows:

The first, which is only a straight line which is length.

The second, which is height, and a shape such as a L or box can be created.

The third, or z-axis, in which shapes with length, width, and heigh can be created.

The fourth dimension is time. You need to know an object's position in time in order to plot its position.

The fifth dimension is presumed to be a world that is slighty different than our own, and would alow us to observe similarities between this world and others. (In my estimation, the fifth dimension is a gateway or portal between worlds.)

The sixth dimension is a plane of multiple possible worlds and universes.

The seventh through tenth dimensions get more complicated, but as this is theory and not known provable fact we will stop here for the purposes of this book.

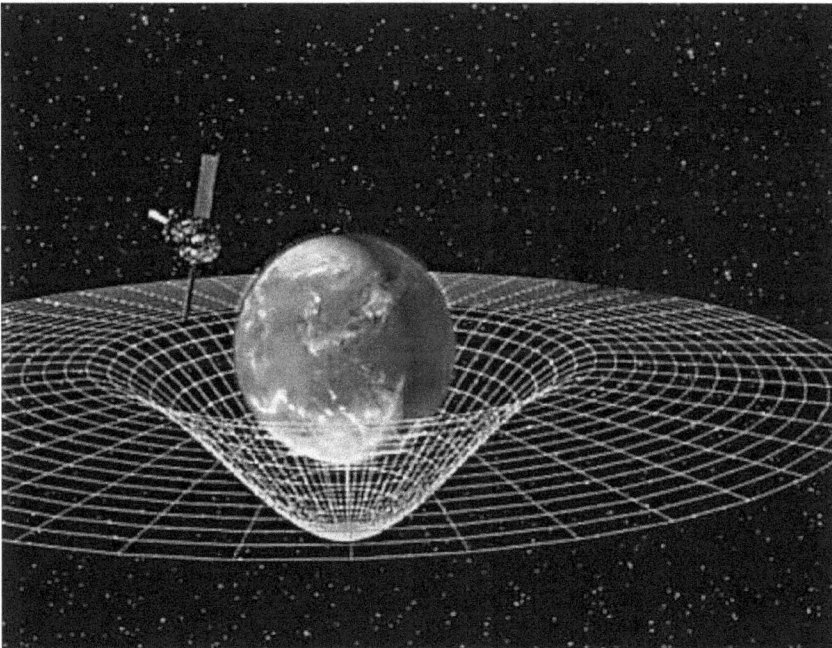

FIGURE 4 ARTIST CONCEPT OF GRAVITY PROBE B ORBITING THE EARTH TO MEASURE SPACE-TIME, A FOUR-DIMENSIONAL DESCRIPTION OF THE UNIVERSE INCLUDING HEIGHT, WIDTH, LENGTH, AND TIME.

The fact that science is considering the existence of other dimensions is interesting, but it is highly likely that scientists have gone far beyond just thinking about the subject and have located and studied natural portals on our planet already, such as the case with the Skinwalker Ranch a the Uinta Basin in Utah. This is a well-known site with extremely strange phenomena occurring on it, and a site where many speculate there is an inter-dimensional portal. The History Channel series "The Secret of Skinwalker Ranch" covers much of the strange phenomena at the site. One such feature is the newly discovered source of gamma rays about one mile above the property, which was a shocking find.

In 2016 Anita (Antarctic Impulsive Transient Antenna) detected anomalous extremely high-energy neutrinos coming up for the Earth's surface but could find no source for this. According to a New Scientist article that *"Seemed impossible."* The article goes on to say that: *"Explaining this signal requires the existence of a topsy-turvy universe created in the same big bang as our own and existing in parallel with it. In this mirror world, positive is negative, left is right, and time runs backwards."*

Tri-field meters used at the Skinwalker Ranch read off the charts at times, which some speculate could be an indication of a portal opening or closing.

In several cases unidentified aerial objects were seen by witnesses just prior to, during, or after they saw a winged creature. In several cases, government agencies have found an interest in this, which prompts the question – Could these creatures actually be aliens from another planet or dimension?

In my opinion, UFO and extraterrestrial, and cryptid researchers need to take a serious and closer look at these types of sightings and consider that these beings may be traveling inter-dimensionally.

In the past, many investigators have dismissed reports of anomalous flying creatures simply because they could not believe the possibility that they are real, even believing that the witness was hallucinating or misidentified a known bird.

However, there are too many credible cases to simply ignore them. We may be dealing with an entirely different type of alien being than has as of yet been imagined. These creatures may be different types or species of winged aliens which are occasionally visiting our third dimensional space where we catch a glimpse of them for a short period of time before they exit our dimension and return to theirs.

Map of Sightings

I created a Google map of giant bird, pterodactyl, fairies, Mothman, and other winged creature sightings. This is an on-going project, and the reader may see updates at this site.

Go to:

https://tinyurl.com/4aw9uucs

Or visit www.margiekay.com to see the map.

If you have had a sighting please report it to me at www.margiekay.com or email margiekay06 @yahoo.com.

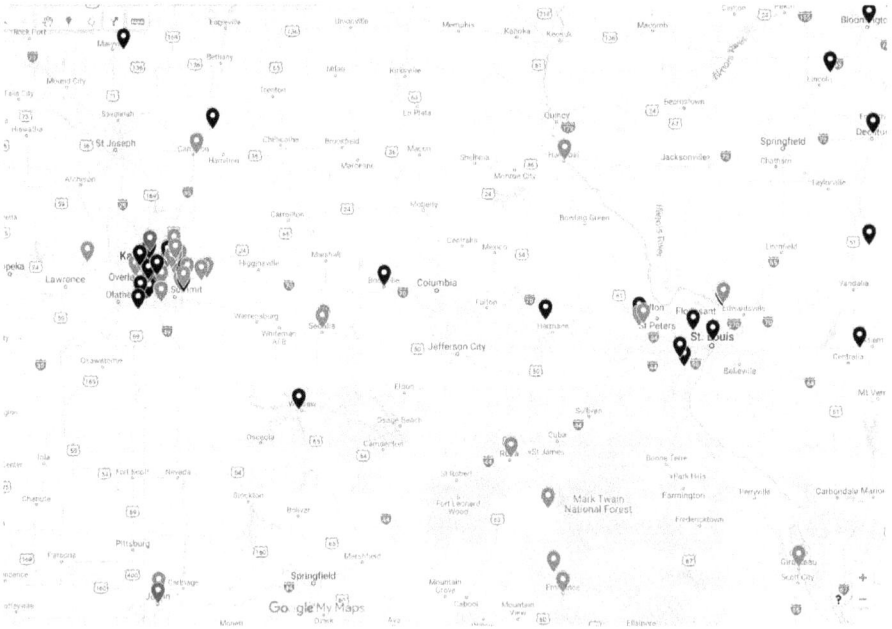

MAP OF WINGED CREATURE SIGHTINGS

GLOSSARY OF TERMS

Batman: The name some people have given to a creature that looks part man/part bat, usually with a humanoid body and bat-like wings with no feathers. The size is usually reported to be 6' – 8' in height.

Batsquatch: A term used to describe a creature that looks like a Sasquatch with bat-like wings. It is primarily in southeast Missouri. The size is usually reported to be 7' – 10' in height.

Butterfly People: A name given to humanoid creatures of varying sizes by children in Joplin, Missouri who were saved by these unknown beings. Some speculate that they were angels, or fairies, and others butterflies with humanoid bodies. These creatures spoke telepathically to the children. The stories remained the same even though the children did not know each other.

Dimensions: Theoretically, there are three dimensions in our physical world, a fourth dimension of time, and dimensions beyond that which so far have not yet been proven by science, but that observers have seen. A theoretical superstring theory requires 10 spacetime dimensions, and the Supergravity theory promotes 11 dimensions so spacetime.

Fairy: A small creature ranging in size from ½" to 12" in length that look like a humanoid with wings. Descriptions vary from the creatures looking exactly like tiny humans to humanoid-like creatures. The number of wings vary from two to four. These creatures have been seen world-wide.

Flying Humanoid: Usually observed in Mexico, these creatures appear human-like but do not have wings.

Giant Bird: A term used when people cannot think of anything else to call a gigantic creature with wings. Sometimes these appear as normal birds, others do not. Wing sizes range anywhere from 10' to 85' in width.

Humanoid: Term used to describe something that looks similar to a human with a body, two arms, two legs, and head, but is not human.

Inter-dimensional: The ability of a being or craft to move between dimensions i.e., from the fifth dimension to the third dimension and vice-versa.

Mimic: The name given to 6" to 10" fairy-like creatures by a couple in Oklahoma. The fairies they see are similar to humans but look somewhat insect-like as well. They all have two human-like legs.

Morph: To change the form or character of. Often used to describe some creatures who appear as one thing then change to another.

Mothman: The name given to a humanoid creature with wings measuring up to 9 feet in height by people in Point Pleasant, West Virginia. Most people report seeing red glowing eyes. The name stuck and people who have seen similar creatures in other locations call it by the same name.

Portal: A theoretical doorway to or from another dimension by which beings pass through. It is theorized that some creatures such as unidentified winged creatures may come into our third dimension via a portal, then return from whence they came.

Pterodactyl: A creature known to have existed millions of years ago. Many people swear they have seen a similar creature in recent times. Usually seen in remote areas near large bodies of water.

Pterosaur: Another name for a Pterodactyl.

Thunderbird: An enormous bird that creates lightning bolts from its beak, and also thunder by flapping its wings. It is the guardian of the upper world in Native American mythology. Some people claim to have seen thunderbirds. They may be supernatural or inter-dimensional beings.

Unidentified Flying Object: (UFO), aka Unidentified Aerial Object. An unknown flying object of any type that cannot be identified as any known aircraft. In some cases winged humanoids have been seen at the same time as a UFO or actually morphed into a UFO.

Vortex: A natural or artificially created spiral vortex which may open a portal.

Winged Humanoid: A humanoid type being that has wings. Usually described as tall with bat-like wings at six to nine feet tall.

BIBLIOGRAPY

Starlink Satellites:
www.find starlink.com: Starlink satellite formations

Video of Starlink:
www.space.com/spacex-starlink-satellites-spotted-night-sky-video.html

Worldwide MothMan and Winged Creatures Sightings Facebook Group:
https://www.facebook.com/groups/WorldwideMothMan

Jessica Schaefer *to* Haunted Missouri: Real stories of the paranormal in the Show Me State

David Hatcher Childress www.s8int.com/dino21.html show - Me-State

The Crypto Crew: www.thecryptocrew.com

Joshua P. Warren podcast on February 18, 2018:
http://www.buzzsprout.com/127013/645529-joshua-p-warren-daily-when-mothman-flies-over-your-house

Paranormal Love to Know:
www.paranormal.lovetoknow.com/urban-legends/recent-mothman-sightings

Portals:
https://www.ghosthuntingtheories.com/2020/05/skinwalker-ranch-secret-portal-to-other.html

Mysterious Radio:
www.mysteriousradio.com/mysteriousradio-blog/2018/8/16/the-mothman

Singular Fortean:
www.singularfortean.com/news/2020/7/29/woman-reports-2002-sighting-of-gargoyle-with-feathered-wings-in-babcock-wisconsin

https://www.singularfortean.com/news/2020/8/5/illinois-man-recounts-sighting-of-winged-creature-with-the-body-of-a-large-man-seen-while-hunting

Cryptozoology News:
www.cryptozoologynews.com/pterodactyl-in-the-corn-fields-of-missouri-town/

NPS.Gov: www.nps.gov/ozar/blue-spring.htm

Giant Thunderbird Returns: www.liveabout.com/the-giant-thunderbird-returns-3862215

Live Pterosaurs: www.livepterosaurs.com/inamerica/blog/

The Butterfly People:
www.realparanormalexperiences.com/the-joplin-tornado-butterfly-people

Only in Your State: www.onlyinyourstate.com/kansas/4-cryptids-in-kansas/

Atlas Obscura: www.atlasobscura.com/articles/the-mythic-child-stealing-thunderbirds-of-illinois

Mt. Vernon Register-News August 4, 1977

StrangeStrangeStrange:
strangestrangestrange.com/paranormal/van-meter-visitor/

The Bigfoot Diaries: www.the bigfootdiaries.blogspot.com

www.s8int.com/dino21.html

Wikipeda Cryptids: www.cryptids.wikia.com

Wired.com: www.wired.com/2014/08/realistic-robo-hawks-designed-to-fly-around-and-terrorize-real-birds/

Live Pterosaur.com:
www.livepterosaur.com/LP_Blog/archives/tag/kansa

St. Louis Today: www.stltoday.com/news/local/metro/the-butterfly-people-of-joplin/article_cca48b1a-282b-587d-902b-cd5f09ca8516.html

The Mothman:
www.themothman.fandom.com/wiki/Winged_Humanoid_Sighting_Reported_After_Minnesota_Bridge_Collapse_2007

The Mothman Prophecies, John Keel, 1975, Signet Books, New York.

Curious Cryptids:
www.curiouscryptids.tumblr.com/post/167247170245/mothman-i35-w-bridge-collapse-on-august-1-2007

Science Rumors: www.science-rumors.com/top-10-mothman-sightings-with-pictures-proved-it-is-real/

The Van Meter Visitor: A True and Mysterious Encounter with the unknown by Lewis, Voss, and Nelson

ECETI Ranch/James Gilliland: www.eceti.org

Visit Cryptoville: www.vistcryptoville.com

Mothman Museum and Festival: www.mothman museum.com

Thunderbird video by Chief A-J: tinyurl.com/2kuynk4z

Cryptomundo: www.cryptomundo.com

John Hyatt Captures Fairies on film: https://tinyurl.com/

Superstring Theory: https://stillunfold.com/science/10-dimensions-of-reality-according-to-superstring-theory

themothman.fandom.com/wiki/26_Unofficial_Mothman_Sightings

Mothman Lives: mothmanlives.com/mothman-sighting-reports.html

Akron Ohio Moms: akronohiomoms.com/ohio/monsters-sightings

The Vintage News:
thevintagenews.com/2019/01/07/mothman

Article in Point Pleasant Register:
http://www.westva.net/mothman/1966-11-16.htm

Rense.com: https://rense.com/general66/humsan.htm

INDEX

ABOUT THE AUTHOR

Margie Kay is a veteran UFO and Paranormal investigator. She is the Assistant State Director for Missouri MUFON and the Director of Quest Paranormal Investigations. She has completed over 1,200 UFO investigations to date. Kay is the president and CEO of four businesses which include a fire investigation company, a chimney contracting business, a publishing company, and a real estate investment business in Kansas City, Missouri. She is a Missouri licensed private investigator.

Kay is an accomplished psychic and remote viewer, having helped to solve over 60 homicide, missing person, and theft cases for law enforcement and private investigators. She has completed over 3,000 private readings for individuals and demonstrates her unique abilities at conferences and meetings.

Margie is the author of 14 books and has been a guest on over 200 radio programs. She appeared on the pilot TV show, Strange; Hangar 1; Ancient Aliens, CNN, and others. Kay has presented at over 200 conventions and meetings nationwide, and she does virtual remote viewing training.

Margie investigates strange phenomena all over the world and is interested in pursuing paranormal events of any type including Sasquatch, haunted sites, UFOs, Aliens, interdimensional beings, time anomalies, and cryptids.

Margie is the host of the Un-X News Radio Show and podcast on YouTube.

Contact:
Website: www.margiekay.com
Email: Margiekay06@yahoo.com

PUBLICATIONS BY UN-X MEDIA

Haunted Independence Missouri by Margie Kay 2013 & 2016

Gateway to the Dead: A Ghost Hunter's Field Guide
by Margie Kay 2016

Family Secrets by Jean Walker 2017

The Kansas City UFO Flaps by Margie Kay 2017

Un-X News Magazine 2011-2016 in print

A Sonoma County Phenomenon: **Evidence for an Interdimensional Gateway** by Margie Kay 2019

The Fast Movers: Evidence for High-Speed UFOs/UAPs
by Margie Kay, Bill Spicer, and Larry Tyree 2020

Journey to Spirit by Devin Listrom 2020

Winged Aliens by Margie Kay 2021

The Master Dowsers Chart Book by Margie Kay 2021

Rules for Goddesses by Margie Kay 1999

Missouri UFO Hot Spot by Missouri MUFON 2021

The Alien Colonization of Earth's Waterways
by Debbie Ziegelmeyer 2021

THOR by Margie Kay 2021

All books available at www.amazon.com

Un-X Media is currently taking book submissions.
We publish non-fiction books about unexplained phenomena.
Please check the website for writer guidelines.

Contact:
editor@unxmedia.com
816-833-1602
www.unxnmedia.com

UNXMEDIA

PUBLISHING

www.ingramcontent.com/pod-product-compliance
Lightning Source LLC
Chambersburg PA
CBHW072144270326
41931CB00010B/1878